The Chicago Guide to College Science Teaching

Chicago Guides to Academic Life

A Student's Guide to Law School
Andrew B. Ayers

What Every Science Student Should Know
Justin L. Bauer

The Chicago Guide to Your Career in Science
Victor A. Bloomfield and
Esam El-Fakahany

The Chicago Handbook for Teachers, Second Edition
Alan Brinkley, Esam El-Fakahany,
Betty Dessants, Michael Flamm,
Charles B. Forcey, Jr., Matthew L.
Ouellett, and Eric Rothschild

The Chicago Guide to Landing a Job in Academic Biology
C. Ray Chandler, Lorne M. Wolfe, and
Daniel E. L. Promislow

The PhDictionary
Herb Childress

Behind the Academic Curtain
Frank F. Furstenberg

The Chicago Guide to Your Academic Career
John A. Goldsmith, John Komlos,
and Penny Schine Gold

How to Succeed in College (While Really Trying)
Jon B. Gould

57 Ways to Screw Up in Grad School
Kevin D. Haggerty and Aaron Doyle

How to Study
Arthur W. Kornhauser

Doing Honest Work in College
Charles Lipson

Succeeding as an International Student in the United States and Canada
Charles Lipson

Off to College
Roger H. Martin

The Thinking Student's Guide to College
Andrew Roberts

Doodling for Academics
Julie Schumacher

The Graduate Advisor Handbook
Bruce M. Shore

The Chicago Guide to College Science Teaching

Terry McGlynn

The University of Chicago Press
Chicago and London

The University of Chicago Press, Chicago 60637
The University of Chicago Press, Ltd., London
© 2020 by The University of Chicago
All rights reserved. No part of this book may be used or reproduced in any manner whatsoever without written permission, except in the case of brief quotations in critical articles and reviews. For more information, contact the University of Chicago Press, 1427 E. 60th St., Chicago, IL 60637.
Published 2020
Printed in the United States of America

29 28 27 26 25 24 23 22 21 20   1 2 3 4 5

ISBN-13: 978-0-226-54222-5 (cloth)
ISBN-13: 978-0-226-54236-2 (paper)
ISBN-13: 978-0-226-54253-9 (e-book)
DOI: https://doi.org/10.7208/chicago/9780226542539.001.0001

Library of Congress Cataloging-in-Publication Data

Names: McGlynn, Terry (Biologist), author.
Title: The Chicago guide to college science teaching / Terry McGlynn.
Other titles: Chicago guides to academic life.
Description: Chicago ; London : The University of Chicago Press, 2020. | Series: Chicago guides to academic life | Includes bibliographical references and index.
Identifiers: LCCN 2019058080 | ISBN 9780226542225 (cloth) | ISBN 9780226542362 (paperback) | ISBN 9780226542539 (ebook)
Subjects: LCSH: Science—Study and teaching (Higher)
Classification: LCC Q181 .M26 2020 | DDC 507.1/1—dc23
LC record available at https://lccn.loc.gov/2019058080

♾ This paper meets the requirements of ANSI/NISO Z39.48–1992 (Permanence of Paper).

For Amelia and Bruce

Contents

Preface ix

1. Before You Meet Your Students 1
2. The Syllabus 26
3. The Curriculum 49
4. Teaching Methods 70
5. Assignments 105
6. Exams 117
7. Common Problems 133
8. Online Teaching 162
    Afterword 178

Acknowledgments 181
Notes 183
Suggested Readings 191
Index 193

Preface

Higher education is a strange beast. Our institutions hire highly trained professionals to teach, but most of us have not been trained in teaching. Scholars of education have compiled a mountain of evidence about effective teaching practices in college science classrooms. Many journals are dedicated to publishing peer-reviewed research about science teaching. Yet, by the time we finish graduate school, many of us feel overwhelmed at the prospect of diving into books and articles about pedagogy, even if teaching is a substantial part of our career. After all, don't we have enough trouble keeping up with the latest work in our own disciplines? A chasm separates scientists from scholars of teaching and learning, and few science instructors have invested in building a way across. I'm hoping this book can help more scientists develop their own bridges.

When I said I was writing a book about college science teaching, one education professional asked me, "Which theorists are you using?" And my answer was "Well, it's not that kind of book." If you can rattle a list of your favorite educational theorists off the top of your head, then this is probably not the book for you. I'll be discussing teaching practices that are rooted in theory, but I'm not here to teach (much) theory. As a research biologist and a classroom instructor, that's outside my wheelhouse.

Scientists have jargon, and scholars of teaching and learning have just as much jargon. We often talk past one another. It is unfortunate that most academic literature is written to be inaccessible to nonexperts. This makes it hard for science instructors and pedagogical experts to communicate with one another. I hope we can have more conversations about what effective teaching looks like and how we can change what we do to help our students find academic success.

The suggestions in this book are grounded in the peer-reviewed literature on teaching. I don't think you need to be an expert in science pedagogical theory to become an excellent classroom teacher. However, it helps to be receptive to what experts have to say and to be prepared to adapt to improve your craft. At the back of this volume, I've provided notes on the academic sources for each chapter. I have done my best to cite large meta-analyses and review papers along with books, which I think will be of greatest utility for a newcomer. I also suggest some additional books, in case you get jazzed about the scholarship of teaching and learning.

Teaching is a practical act. When we step into the classroom, we are not clouds of theory. We are actual people, supporting the learning of others. In this book, I suggest a range of actions that you can take in the classroom, which are informed by research, theory, and experience. Ultimately, your choices can help you build an impactful relationship to support the success of your students.

If you're new to teaching at the university level, and you have some time before the start of the semester, then you have the opportunity to digest this guide from the start. But if your teaching assignment starts very promptly, I suggest jumping straight to the syllabus chapter and then going to associated chapters on curriculum and teaching methods as the need arises. If you're more seasoned in the classroom, then I am hoping that I might be able to provide some fresh perspectives, to nudge you into changing up some aspects of your teaching. I think the chapters most of interest to you will be chapters 1, 4, and 6. If you're a graduate student teaching lab sections, chapter 4 on teaching methods will be of particular interest, and the tail end of the chapter is specific to laboratories. If you are teaching an online course, the chapter about online teaching will clearly be of particular interest, but the entire book is broadly applicable to online learning, aside from some parts of chapter 4.

This book is intended for instructors in higher education in STEM from around the world. After all, effective teaching practices work wherever you are. Nevertheless, my experience is primarily in the United States, and our system of higher education has some unique features. I aimed to be as inclusive as possible without compromising the points I was making, and to acknowledge distinctions when necessary. You may have already noticed one cultural difference: the title of the book. Here, "going to college" is synonymous with "attending university," and the term "college" exclusively refers to postsecondary education. (Two-year institutions and some smaller four-year institutions refer to themselves as "colleges," and universities are also composed of smaller units that are called colleges. It's rather confusing.) I've attempted to avoid confusion by choosing terminology that is more generalized. For example, the terms "professor," "lecturer," and "teacher" have different meanings or connotations around the world, so referring to us as "instructors" is more universal.

I also have aimed to be inclusive of all STEM-related academic disciplines. Teaching is teaching, whether you're a physicist, geographer, or computer scientist. That said, I recognize that there are distinct practices and challenges associated with teaching in every discipline. As a biologist, I'm experienced with running a biology classroom, but I've only been a student in other science disciplines. I've taken steps to make sure that my book will speak to you, regardless of what kind of scientist you are. The book is periodically peppered with remarks from scientists from a wide variety of academic fields and types of institutions. In the preparation, I interviewed dozens of instructors about how they run their classrooms, to gain a better idea of the range of practices. I chose to limit the number of examples featuring content from a specific field. Sometimes, however, a concept is best communicated when illustrated with a concrete example. The examples are designed to be general enough that most of us would be familiar with them from our prior education.

What we do in our classrooms matters more than ever. Humankind is deep in the climate crisis, and our way out requires an educated populace. For more than a billion years, our planet has been pulling carbon from the atmosphere via photosynthesis and banking this carbon underground. We have broken into this bank to spew carbon back into the atmosphere at a terrifying rate. We have all of the technology necessary to address this problem but are lacking the collective will. To make it worse, a substantial minority of people don't accept the straightforward science of this problem. To be a science educator at this moment is consequential. Teaching science won't directly stop carbon pollution, but nonetheless, we are ambassadors of science. We are the ones who are most responsible for putting a human face on science. Some of our students will become scientists, and it would serve us well if we filled our teaching with wisdom from beyond the sciences. As we teach nonscientists, we must be inclusive, to show that science is for everyone and that everyone needs science. Understanding science matters, but this only makes a difference if this understanding is leveraged to create a better community for all of us.

1

Before You Meet Your Students

Teaching Is a Choose-Your-Own-Adventure

Teaching is a series of decisions. These decisions structure how well you achieve your goals and how much everybody enjoys the experience. As you are planning a course, picture in your mind an intricate flowchart, with thousands of options. What books will you use? Are you going to have weekly quizzes? Do you give students your lecture slides in advance? Do you accept late assignments? What questions will you put on the exam?

This practical guide isn't designed to tell you which turn to make at every decision point. My goal here is to provide you with some perspective so that when you come to these forks in the road, you can have a better idea how that choice might affect the route ahead. We're all driving in different places, and our classes have different destinations, so I can't just prescribe a particular route! We all need to learn how to read the map, the educational landscape of our classroom, and do our own wayfinding.

This book is constructed with two keystones: Efficient Teaching, and the Respect Principle. I picture these as keystone concepts rather than foundational concepts. The foundations of teaching are your expertise, your time, and your motivation. That's all you need to break ground for the semester. A keystone is the block at the center of an arch, holding up a structure that has been assembled. Once you build your course, it's the keystones that will prevent your work from crumbling in a big heap in the middle of the semester. When you respect your students, and you choose teaching approaches that have the best educational bang for the buck, you can help create learning that will stick with students well after your course has ended.

Efficient Teaching

Teaching well takes time and effort. There's no magic shortcut. This introduces a dilemma: The more time you invest, the better it is. Every instructor is compelled to identify the moment to stop putting work into a course, by constraint or by choice. The threshold when we have invested enough time into teaching, and what we have done, needs to be good enough to meet our standards. This decision—when to stop preparing for class—is only one variable that affects how well our students learn.

For most of us, there are many factors that prevent our classes from being perfect. A lack of time investment might not be the biggest shortcoming. Many of the shortcomings won't be overcome by putting in more work, because we still have much to learn about teaching effectively, and the classroom environment is always changing. We can improve our craft, to a large extent, by making effective decisions and choosing approaches that result in student success. The principle of teaching well without letting it consume your whole life is what I call "efficient teaching."

When you teach efficiently, the quality of student outcomes is high relative to the amount of time that you invest. An efficient teacher is capable of delivering a great course without breaking into a sweat, or staying up late on a regular basis to get caught up with teaching duties.

Trying to perfect your course by sheer effort is inefficient, and attempts to do so will grind you into exhaustion. Just as there's no magical shortcut to excellent teaching, there also is no shortcut to magical learning. Because students only have so many hours in the day to focus on the content of your course, you can only spend so much time structuring the template for their learning. Of course we shouldn't blow off the courses we are teaching, but when we make an investment, it should be designed so that our work translates into more investment made by students. There's no use teaching your butt off unless the students learn. Students don't learn as a consequence of your performance; they learn when they work hard and they are personally invested. A well-designed course will see to it that your students' efforts pay off in their learning.

How do you invest your preparation and teaching time to garner the biggest bang for the buck? That is your personal choose-your-own-adventure. The time invested into teaching produces diminishing returns. If you've already put five hours into preparing a lesson, is another hour going to make much more of a difference? There's no need to pour effort into your class if it only causes a tiny difference in student outcomes. If you teach efficiently, then you can hit that point of diminishing returns more promptly.

How much time you invest is not as important as making sure that you are making the most effective choices about what you are doing in the classroom. For example, if you're teaching organic chemistry, you could spend a whole day writing and producing a three-minute music video about covalent bonding to the tune of a pop song. It would definitely be weird and memorable for your students, but also a massive time sink for you as an instructor. Doing all of this work on your own, just for a relatively small level of engagement by the students, is inefficient. If you think that making music videos can help students learn, then you could assign such a thing to your students.

What affects the relationship between your effort and student learning? This is a complex question with a complex answer, which differs with every instructor, course, and cohort of students. This is a question to have in mind as you're prepping your course. It's helpful to ask yourself, "How exactly is this going to help my students learn?"

Some highly effective teaching practices are inherently inefficient. A classic example is editing multiple drafts of student writing. Instructor feedback helps develop critical thinking as well as writing skills, but it also takes a lot of your time. Likewise, frequent assessments can improve engagement and promote effective study habits. However, developing and grading all of those assignments will take time. This doesn't mean that we should avoid writing assignments or avoid frequent assessments, because they are impactful. We should be cognizant of this trade-off, so we can maximize student gain given the resources available to us.

The good news is that there are a variety of highly effective practices that are time efficient. Some examples of these are case studies, frequent

...ks on understanding during class, questions that allow students to interact with their neighbors in class, or ungraded snap quizzes at the start of every class. Some technology-intensive approaches can be effective without much additional time (clickers, moderated question boards), while others may take more time to set up than the payoff is worth (quizzes in the course management system, creating online videos to supplement in-class instruction). We'll explore teaching methods in detail in chapter 4, to help you find what works best for you and your students.

Because we all have different experiences and aptitudes, what is efficient for one instructor may not be for another. For example, some instructors report that certain types of active learning can take a substantial amount of their effort without improving student outcomes. Others see a big payoff without much increase in instructor effort. It's your class, so you've got to figure out what is efficient for you.

The Respect Principle

The Respect Principle is as simple as it sounds: Teaching is more effective when we respect our students. Whenever we make one of the many thousands of decisions in the course of our teaching, we should be asking ourselves, "Is this respectful of our students?"

What does respect for students look like? It means trusting students. It definitely helps to like them as fellow human beings. It means accepting them as they are. It means realizing they might not have as much experience, but that their perspectives have value to the community.

At first glance, the Respect Principle sounds simple and unobjectionable. Of course we all should respect our students! That said, once you wade into the trenches of a semester, I think there are a lot of challenges and barriers to building and maintaining a respectful relationship with our students. Many of these challenges emerge from a milieu in higher education that emphasizes adversarial relationships with students over a spirit of collaboration.

If we teach the way that most of us were taught, then we are going to be doing a lot of things to tacitly communicate to students that they are not worthy of our respect. Sometimes it's just the little things, like how we might communicate that our time is more valuable than their time. In many quarters, it's common for instructors to speak disparagingly about point-seeking behaviors of students, or to discuss how students have fabricated excuses. On a broader scale, many of our policies disrespect our students' agency by micromanaging their time and by implementing grading schemes that are more punitive of undesired behaviors than evaluative of academic performance.

To treat our students with respect, we might need to undertake a large-scale rethinking about how we operate our courses: how we create and grade assignments, how we deal with attendance, how we communicate what goes on a test, and how we teach every day. Respect requires us to think about the classroom experience from the student perspective and to create an environment that best supports learning. Having respect for our students means that we have to reconsider our own unsupported assumptions about what constitutes effective teaching, because our students deserve a quality education, built on evidence about effective teaching practices. A lot of college instructors have very firm ideas about what works best for them in the classroom. In contrast, respecting our students means we need to stay open to the idea of doing things differently, if we are to learn how to help our students learn better.

Some students walk into the classroom on the first day eager to learn, but I suspect more walk in with a measure of fear and cynicism. Considering how some professors behave, how can students not be afraid? Every student has been burned at some point in their academic career. They might have had their grade knocked down as the result of a capricious decision by one of their instructors or by an instructor who incorrectly assumed that they weren't being honest. Unless students enter your classroom knowing that you have a reputation for being a kind and fair person, they'll be cautious that you'll turn out to be someone who doesn't respect where they're coming from. In the eyes of many students, their instructors treat them as if they are guilty until proven innocent. Student work might be automatically screened for plagiarism. Students are too often suspected of trying to get away with something. If a student tells their professor that a member of their family died and they need time to grieve or to travel, this might be met with the default assumption that they are lying. If a student comes in to show concern that their exam was misgraded, the professor may assume the student is asking for something they didn't earn. If a student says they're sick, then they might be required to prove it.

Our students aren't out to get us. When students walk into our classrooms, they are not sure if their professor is expecting an adversarial relationship. They don't necessarily feel trusted. Even if you don't think this about your own students, we—collectively in the professoriate—are part of this system that treats students as inherently untrustworthy. We do not have instant access to the trust of our students because all students have had adverse interactions with prior instructors, and they would be wise to be wary.

"Have respect for your students" is a simple phrase, but following through isn't always straightforward, because the culture of higher education is steeped in conventions and philosophies that undermine a respectful relationship between instructors and students. Building a respectful relationship with students takes more than just treating our students like fellow human beings. To allow mutual respect to flourish, we need to intentionally dismantle the standard features of the college science classroom that are built on distrust. Because instructors and college students often are lacking the environment to support a healthful learning relationship, it's our job as instructors to reimagine what our relationship should be like with our students and to create the conditions that allow it to emerge.

What are some features of treating college students with respect? There are two aspects of the Respect Principle that I'd like to explore in more detail: respecting the complexity of the lives of students, and treating our students fairly.

Students Have Complex Lives

It's been a tradition for scientists to run their courses without choosing to fuss about student responsibilities beyond the classroom. I suspect this attitude has two origins. First, college professors disproportionately come from families that send fresh high school graduates to enroll in college, where they will expect to focus well on their studies and then graduate within four years. Since many professors were capable of having a laser-like focus on academics while they were undergraduates themselves, I suppose it's easy to have the same attitude toward their own students.

Second, the ethos of higher education is still stuck in picturing college as a bucolic idyll distinct from the realities of family life, employment, and the broader social and political environment. I think there has been some progress in recent years, but nonetheless, many instructors continue to consider that being in college is somehow different from living in the "real world." It is hard for students when their professors don't recognize that they have lives that extend beyond their academic pursuits.

Should we have high expectations of our students? Definitely! Should we design our courses so that it's impossible to succeed without devoting one's life to it? I think not. We very well may have experienced that inspiring professor who set the bar so high that the only way to get an A in their course was to devote every breathing hour to the topic. However, applying the Respect Principle will mean that all students in the course should have the chance to do very well in the course if they invest a reasonable amount of time into the work outside class.

What is a reasonable amount of time that we can expect from our students outside class? I'm not sure what constitutes "reasonable." But I can tell you what appears to be a frequently expressed behavioral norm for students: three hours of work for every contact hour. If your class meets for three hours each week, then this would mean students are expected to spend nine hours outside the class per week. If we expand that to students who are enrolled in a typical full course load of 15 units per semester, that would mean students are spending 15 hours with their butts in seats for coursework and 45 additional hours working on these courses outside class, including reading, problem-solving, projects, homework, term papers, and so on.

Is it reasonable to expect full-time students to spend 60 hours per week on academic work? Including weekends, that's not much more than eight hours per day, which on the face of it seems somewhat reasonable, right? I think this is reasonable if we make the assumption that students don't have much going on in their lives other than being in college. However, that's often not a safe assumption. How many hours outside class can we expect of our students? That's a difficult question to answer, and how you find your own answer will depend on the norms of your institution, the course that you're teaching, and how you've designed the course.

Students attend college because their education is part of the real world. In the United States, about two-thirds of students are commuters, and the majority of students are working at least 20 hours per week. These jobs might be as a tutor or a lab tech, but more often students are doing work unrelated to their academic pursuits, such as tending bar, delivering packages, giving campus tours, or working in retail. For example, I am working with a student who washes dishes several nights per week, and another works the night shift in a large pet shop. Both of them are also taking a full course load. If you go through a student parking lot on campus, do you see stickers on student cars for ride-sharing services? Even though this comes as a surprise to many faculty members, many college students are living on a financial edge.

Today's college students are experiencing different financial circumstances than most instructors. Things are definitely not the way they used to be, even as little as five years ago. While real income for most families has not grown over the past decades, the price of attending college has increased exponentially. Not long ago, it was common for universities to provide financial aid packages to students that would meet all of their demonstrated financial need, while keeping loans at a manageable level with a reasonable expectation that they could be paid off when they found a decent-paying job. That is extremely rare now.

The days of affordable college are long behind us, and it might be worse than you think. The average amount of educational debt per college graduate is $30,000. Odds are that many of the students that you are teaching are going into massive levels of debt just to be in college, or they are working many hours to avoid this debt. This is not something that you can necessarily know about students just by looking at them, or by seeing if they are wearing designer clothes or have what appears to be an expensive book bag.

Students in tight financial straits are not just exhausted from schoolwork and employment; they might be experiencing shortcomings with respect to their basic needs. If a person doesn't have a safe and consistent place to sleep every night, and if they are unable to eat healthfully on a regular basis, this obviously is a detriment to their well-being as well as their academic performance. In the nation's largest university system, the California State University system, an internal study found that about one-quarter of all students are food insecure, and one-tenth of students meet the criteria for homelessness. This is also a common phenomenon in high-tuition and high-endowment institutions, many of which have recently established food pantries on campus to meet the needs of their own students. You might have students who are sleeping in their cars, or surfing couches of acquaintances, and students taking exams in the morning even though they were working the night shift and weren't able to eat a reasonable breakfast. And it's impossible to know whether any particular student is struggling with these difficulties, because this is rarely something that students will readily share with their instructors, and there is great social pressure to disguise this level of financial insecurity.

Financial precariousness is not endemic to universities that predominantly serve low-income student populations. Even if you are teaching in a well-endowed private college where the student parking lot is filled with cars that you could never afford, please know that there are some students in your midst who are barely scraping by. You won't be able to identify these students by their clothes, their demeanor, or their accent. Their lack of security is typically invisible, and their challenges may even be amplified by attending school in an environment of conspicuous wealth.

Showing concern for the welfare and private lives of students isn't just something that low-income students deserve. All students—including the most-privileged ones—have complex lives that go far beyond the courses that we are teaching. They have professional ambitions that extend beyond their college education, involving internships, or work, or other extracurricular activities. Student athletes are busy with training and competitions. Some may be caring for family members. Many students have chronic illnesses or other disabilities that we might not be aware of, which prevent them from spending all of their time on academic work. It would be remiss of us to fail to recognize that college students are doing more than just taking classes in college.

The Respect Principle can inform how we interact with our students in light of their complex lives. We can develop course policies that seek to accommodate students, in a way that does not marginalize them or put them at a disadvantage to other students. We can choose to not waste their time with busy work that isn't connected to meaningful learning. We can make the most of our time together in the classroom by making sure that lessons are not just for the transmission of information, but for the process of learning. This is discussed in chapter 4.

Treating Students Fairly

We should be fair. This sounds easy enough, right? Would that it were so simple. Among student critiques of adverse experiences in college classes, "The professor wasn't fair" is perhaps the most common. Yet, if we ask these instructors if they treated all students fairly, they will report, "Of course I treated all students fairly." Let's respect our peers enough to acknowledge that they think they are running a fair classroom. How can we reconcile this discrepancy?

This type of conflict emerges from a lack of mutual respect. If a student is being treated fairly, yet they make an accusation that they are not being treated fairly, that comes from a mindset where they do not have respect for the professional integrity of their instructor. If the student's complaint about unfair treatment has some degree of legitimacy, that is because of an error by the instructor, or it ultimately may come from a lack of respect on the part of the instructor. The pathway to reducing the incidence and perception of unfairness is to build relationships of mutual respect.

Sometimes students harbor a general concern that an instructor is not treating them fairly, based on vaguely graded assignments and observed interactions with others. They might be right, or they might not be. In these cases, I think the germane concern isn't sussing out who was right and who was wrong, but instead, asking what disrupted the trust and respect between the student and the instructor. This perceived unfairness makes for a less productive learning environment. Even if the professor is being fully fair, if students don't trust that they are being treated properly, they'll have more difficulty learning. Instructors who are interested in providing opportunities for all students to learn should take steps to build trust with their students.

Instructors may have developed policy specifically to be fair to students, and yet some students might regard that policy as unfair. For example, an instructor might not accept late assignments and might say that they have chosen this policy with the interest of being fair to students who do their work properly and on time. Students who experience an adverse circumstance beyond their control that prevents them from turning work in on time might complain that this no-tolerance policy on late assignments is unfair. Is it fair to a student if their hard drive crashed or they were dealing with a personal crisis, and this prevented them from turning an assignment in on time? In this situation, what constitutes fairness is in the eye of the beholder. To some, treating all students equally is a hallmark of fairness. To others, equal treatment may result in substantially unfair outcomes.

I suggest that trust is undermined when equal treatment is designed in a way that is inequitable. There is a mountain of difference between equality and equity. Equality means that everybody gets the same treatment, whereas in an equitable classroom, everybody has the same opportunity to succeed regardless of their circumstances. If every student is being treated equally, then it's possible that many will be treated inequitably. For example, if you write tests that take a lot of time for students to complete and you collect exams before everybody finishes, then you are functionally providing an advantage to the students who take tests more quickly. Unless the purpose of your class is to train students with the ability to take tests quickly, this is unfair. It's equal, but inequitable. Likewise, if a student's participation grade is determined by how often they raise their hand in class, then this is inequitable to introverts. If your primary mode of testing is using multiple-choice tests, keep in mind that some students have backgrounds where they've received a lot more preparation to excel on multiple-choice questions than other students. And so on.

Considering that there is some element of inequity to many standard features of a college-level science class, I'm not sure it's even possible to run a college course that is 100% equitable. As instructors, we all reckon with practical constraints. We all have our blind spots, and we can never know everything that each of our students is going through. Nevertheless, while 100% equity may not be achievable, that does not liberate us from the responsibility of placing equity as a priority and a key criterion in our decision-making process. If we conceive of teaching as a choose-your-own-adventure, then the Respect Principle requires us to embed equity at every juncture.

We owe it to our students to have high expectations for their performance. The Respect Principle also means that we should make sure that our expectations are clearly communicated and are reasonable. Please keep in mind that we all have had college professors who have not treated us this way. If we strive to teach the way we have been taught, this could turn out rather badly. After all, "I had to go through this when I was a student, so it's only fair that you have to go through this" is tantamount to hazing. Our lens on the college experience is probably different from our own students', so what we perceive as reasonable might be considered to be unreasonable by current students. Many of my suggestions in the next chapter are about establishing clear and fair expectations of all of our students.

Students Are Not Our Adversaries

The college experience is often treated as a game, with instructors and students as adversaries. In this scenario, the instructor is the gatekeeper who places obstacles that students must surmount or circumvent. Inside and outside the classroom, we will always have some people in our lives who will want to treat our interactions as a game to be won. You don't have to be held captive to an adversarial mindset, just because a fraction of the students in your class hold this mindset. If we respond to skulduggery by letting students turn us into the referee, we've spoiled the educational experience for everybody.

When you're running your own classroom, you can opt out of the game.

You can choose the role that you wish to play and build an classroom environment befitting that role. Even if you have some students who are primarily focused on gaming the system, it's still your system. If the system is primarily built to foil those who are trying to game it, it's not going to be optimized for the students who are working earnestly. We must take account the reality that there will be students gaming the system, but if we design every aspect of our courses around this minority of students, then we have ceded our control over the learning process. It is possible to create a classroom environment that prioritizes and values learning and is designed to support students rather than constrain and evaluate them.

What's the best way to conceive of the faculty-student relationship? Students are responsible for their own learning, and the success of our teaching is contingent on the quantity and quality of their work. There is a huge power differential between college instructor and student, as we are not just evaluating the student, but are fully in charge of developing the criteria by which students are evaluated. When we interact with students on a daily basis, how we conceive this role affects how we treat students and how they feel about us and the course.

While running a course, instructors will build a relationship with the students. You can let this happen subconsciously, or you can intentionally create the relationship that will be most impactful. I've recommended against becoming a referee of the academic game. In lieu, some adopt the role of a family member, such as an aunt or uncle. (I often hear professors even refer to their adult students as "kids.") Others think of themselves as a supervisor, with the students serving as employees. An alternative model is to serve as a dispassionate journalist reporting the facts, or to serve as an enthusiastic evangelist for the discipline. Some instructors run their courses as if they are commanding officers in the military, or as judges, who serve in judgment over students as defendants. As a student, I've had all of these professors. In my opinion, these approaches bring along too much baggage and subvert the learning process. A lot of students don't want their instructors to act as if they are their aunts or uncles, or their bosses, or their judges.

There is one metaphor that works quite well: <u>Be a coach</u>. This kind of relationship captures many of the essential features of what universities expect out of students and their instructors. Athletes are ultimately responsible for their own achievement, but their coaches should be experts who provide guidance and support. Ultimately, coaches have authority over their domain, but results are achieved through positivity and hard work, and like it or not, in the end a student's ability to perform the tasks prescribed by the coach will reflect your success.

I like the teaching-as-coaching approach because it removes the adversarial flavor. If we approach our students with the idea that we are all on the same side and that we actively want them to succeed, then this best helps the students learn. Which is, after all, what we're there to do.

I use one phrase frequently: "meeting the needs of students." I picture this dynamic in the context of a coaching relationship. When we take steps to make exams less stressful for students, we are meeting their needs. I don't mean that we are supposed to coddle students or meet the unreasonable demands of students who pronounce what they think is best. We are the experts in the room who get to decide what the educational goals are in the classroom. When we meet the needs of students, we are creating an environment that is conducive to learning. While it's not our job to give students what they want to make them happy, we should do our best to teach in a manner that challenges students without making them anxious or miserable. Because that's good for learning.

Developing Your Own Style

There are many routes to becoming an excellent and seasoned teacher. There are also many destinations, as there are many ways of being a great teacher. Teaching styles emerge from individual personalities and talents, and what works for one person might not work for someone else. It takes time to find a style that works for you. As we gain experience and evolve as people, and as the populations we teach are also evolving, we need to continuously adapt our style. Teaching well in the long haul is not a matter of discovering the recipe that works and sticking with it. It happens when you become well versed enough in the kitchen that you can cook a meal without the recipe book.

When new college instructors start running their own courses, they typically rely heavily on their prior experiences in the classroom as students to develop their courses. Think back to your own professors—what did they do in the classroom that you thought was particularly successful? I'm sure you can think of things that some of your professors did that you thought were a waste of your time or counterproductive. If you adopt the good things and drop the bad things, then surely you're on the right path? Well, it might not be that simple.

When I started teaching on my own, I took the best ingredients from the various classes I took as a student and combined them into a single class. The problem with that approach is that these ingredients came from different recipes. I knew I liked the ingredients, but when I put them all together, it didn't taste good at all. I learned pretty quickly that I couldn't just combine my favorite teaching ingredients into a single course. I needed to find complementary ingredients to invent a whole new dish, of my own design.

It would be unwise to merely model your teaching after what you liked about your own experience as a student. We were not typical students. While most folks graduate and then move on to do other things, we chose to go back for a second serving, and now are back for thirds. Even if you didn't picture yourself teaching college students while you were in college, you are now. If you target your teaching methods toward a younger you, you'll be missing a lot of the students in your course.

Good teachers empathize with their students, but your personal experiences can only take you so far. Times have changed—even if you were an undergraduate not so long ago, the students come from a different perspective, and also the norms in classrooms have been evolving rapidly. Moreover, it's unlikely that you're teaching at the same place where you were an undergraduate. Every place is different, and practices that were normal for you as a student might not fit in where you are teaching now.

I recommend against adopting a different persona when you are in the classroom—you need to be yourself, because pretending to be someone else for a whole semester is exhausting. This doesn't mean you need to bare your soul to your students, and it doesn't mean that you behave in class just like you behave outside class. You can simultaneously be yourself, but not demonstrate all facets of yourself. You can be the kind person you are outside the classroom, while acting with authority and without being conciliatory.

You don't want to choose teaching practices that you think will fail, but be prepared for the idea that some great ideas won't work out, and some ideas that you are skeptical about might just be the thing that works. Teaching is always a moving target. You evolve, the students evolve, the content evolves, and our understanding of best practices in education evolves. So, once you develop your teaching approach, know that you've got to change if you're going to stay effective. It's not about being fashionable; it's about not going stale.

Who Is Responsible for Student Learning?

I've worked alongside many faculty who have said, "It's my job to teach and the students' job to learn." That statement, taken literally, merely asserts a self-evident fact. Nevertheless, when I try to understand the intended implication, I gather it's something like "I can decide the best way to teach my own classes, and regardless of what I do, it's the students who are responsible for their learning." I think that's almost a sensible way to think about teaching at the university level, but it falls short.

It should go without saying that students are responsible for their own education. To some extent, of course, instructors are responsible for student learning outcomes. After all, if students emerge from your course and nobody has learned anything at all, you do have to admit, at least to some extent, that you are responsible for this shortcoming. Then again, isn't it possible to perform very well in the classroom and still not have students learn anything, if they aren't motivated to learn? I also hear among educators the cliché "You can lead a horse to water, but you can't make it drink." This, I believe, is a downright counterproductive way to think about teaching.

How students feel about us affects how well they learn. As Neil Gilbert (graduate student, University of Wisconsin) reflected: "When I think back to my undergraduate days, the classes that I enjoyed the most and learned the most from were taught by professors that I respected and genuinely liked. I think it is also important to care about your students' learning—or at least, to fake it really well. They won't care if you don't care." While not all students may need to feel your concern, it can disproportionately influence students who are struggling, says Eduardo Ayala (instructor, Citrus College): "For my low-performing students it matters greatly that they know that I care about their learning. Every semester I have 3–5 that start off doing poorly because they are terrified about the material, and they expect their instructor not to care. Unfortunately, there are instructors that don't feel it is worth their time to invest energy in a student that is struggling with a class. So I try to spend time with students based on their need." Often, merely expressing genuine concern can make a difference.

We may or may not be responsible for how much students learn. However, if we are teaching well, then we provide students with the greatest opportunity to learn. There's a boatload of research showing that students learn when they are engaged and motivated, and that when students are disengaged and unmotivated, little learning happens. While it might not technically be our job to motivate students, if we want them to learn, then we need to make sure to use approaches that are engaging and effectively motivate students. Yes, it's "our job to teach," but if we are any good at it, that means we avoid doing things that disengage students, and we take the time to do things that engage students. It's their job to learn, and by providing an engaging environment, we make that job easier for them and they'll learn more. Which makes us more successful at our jobs.

With Great Power Comes Great Responsibility

Instructors have a huge amount of authority and power at the college level. We are totally untrained about how to fairly wield this power. I think a lot of the pitfalls in teaching—for both new and experienced instructors—emerge from difficulty in managing our power and in establishing a healthy relationship with students that recognizes this power differential.

Effective teachers occupy their power with aplomb and gain student buy-in by being prepared and respectful. Instructors have near-total authority over everything that can and can't happen in their classes. This kind of authority is not normal in other spheres, and it takes experience and grace to use this power well. This broad power is a feature of universities, and students are fully aware that we hold the capacity to be unfair and unreasonable. Frankly, we've all known professors who have been unfair and unreasonable. How students regard professorial authority varies widely among institutions, fields, courses, and the identity of the instructor. Depending on who you are, where you work, and what you are teaching, your teaching should be very different. Building a respectful relationship with our students is critical, as well as what it takes to maintain that relationship.

It's important to understand that instructors have a lot of power. Using this power appropriately is a necessary part of being fair to students. However, it's easy to fall into the trap of becoming too comfortable with one's power. You will come to many points where you need to be decisive, but you should not be capricious.

If you're not comfortable being in charge of a classroom and you're about to teach for the first time, there isn't a formula for growing into this role overnight. It's okay for you to take some time to figure out how to comport yourself. Remember that while it might not feel natural to you to be in charge, students are entirely used to this kind of arrangement.

Challenges Tied to Instructor Identity

Students do not necessarily treat all university instructors with the deference that one might expect. I'm a bald white guy with an actual gray beard, and as this matches the stereotype, it's easy for my students to see me in this role. When I walk into a classroom, students will typically just assume that I'm the guy running the show. This is a combination of my age and ethnicity and other professorly aspects of my identity. I don't even need to earn student respect for my authority; I usually get it unearned, and then it's mine to lose. On the other hand, some instructors start out without the respect of their students, purely on the basis of their identity.

The research is unambiguous that college students are more likely to respect the expertise and authority of gender-conforming white men. On the other hand, faculty members who are members of underrepresented ethnic groups enter the classroom with less unearned respect of their students. The effect of gender and ethnicity identity intersects with age, so that a young-appearing woman of color has a much steeper hill to climb in the classroom. Novice instructors sometimes are concerned that their authority might be undermined by not being well established as an academic expert qualified to teach the course. If a student is challenging your authority to run the class, that's because of their own issues, it's not about you. If students have a question about your training or authority, then this presumably isn't about your training, but instead something else about your identity that challenges the student's worldview.

If any students don't recognize your professional position as the instructor, this is not your fault. Experiences vary with gender, age, ethnicity, and institution type, and this will greatly affect how professors establish and manage authority. As a way to manage his position in the classroom, Prosanta Chakrabarty (associate professor, Louisiana State University) said, "I wear nice clothes, usually slacks and a sport coat, business casual. I'm comfortable like that, and I look kind of young, so I want to be taken seriously. I'm a goofball and if I dress like the [students] they will think I'm soft." Choosing your outfit is more critical for graduate students. Jessie Williamson (PhD student, University of New Mexico) said, "I dress to look decently professional and to feel comfortable standing at the front of a room in front of 20+ students. I do think it matters. It's a job, and I like to look professional enough to be treated with respect." We should be able to earn respect purely on the basis of the role that we fulfill. If we dress in a manner that students perceive is not befitting our role, this might make running the classroom more difficult. This inequity is rooted in biases held by the students. You should not be expected to conform to unreasonable student expectations. Considering that this situation is pervasive and there is no magical fix to this circumstance, there are a number of common strategies that people adopt: dressing more formally, having students refer to them by their proper title, and correcting students when they take liberties with familiarity or when they assume a lack of expertise. For those who do not experience these chronic disadvantages when interacting with students, we can support our peers by not taking the liberties that are only available to us because of our identity, such as dressing more casually. We can also make a point to recognize the expertise and excellence of our peers when the occasion arises.

Finding a Teaching Philosophy

Teaching can be exhausting, especially the first few semesters. That's because making decisions is hard. As we go about our day, thousands of little junctures cause decision fatigue, causing us to make poor decisions. When you adopt a coherent and consistent teaching philosophy, decisions become easier and you're less likely to make bad ones.

I've written this book to provide context as you approach common decision points. I'm not here to prescribe decisions for you—but it helps to know what might happen down the line after making some early decisions. The decisions that you make will come more easily, and will be more consistent and wise, when they are guided by a coherent teaching philosophy that is student centered. Your motivation is your own business, and teaching doesn't have to be what gets you out of bed in the morning. A teaching philosophy, on the other hand, is about what you think will best support student learning. A useful teaching philosophy guides your decisions and provides a direction for what happens in the classroom. When any issues pop up, you can look at your teaching philosophy as a signpost.

A teaching philosophy is your vision for how teaching affects the educational development of your students. It's not about why you are teaching, but about how to teach. In other words, what is your main goal while teaching, and what is the path that students in your course will take?

Let me provide some examples of some teaching philosophies, starting with my own. Early on, I wasn't aware of any particular teaching philosophy. In grad school, I was just in the classroom as part of the package while earning a PhD. In retrospect, I see that I adopted a teaching philosophy without thinking much about it: that people learn science because it's exciting and full of wonder. As a result, I designed my class around how cool science is and how it's amazing that science explains natural phenomena. My approach has evolved over time. I'm interested in improving critical thinking skills and scientific literacy. At this writing, my teaching philosophy is "People don't truly learn something unless they discover it for themselves." This guides me as I develop my syllabus, assign homework, put together lessons, and deal with issues that students bring to me.

When I asked Helen McCreery (postdoctoral fellow, Harvard University) about her teaching philosophy, she said, "Education should be data driven. We're getting more data all the time about teaching environments that lead to the largest gains in student understanding; we should use those data. One data-driven takeaway for me that I think a lot about is creating a supportive environment. I want all students to know that I am invested in their learning and on their team." Prosanta Chakrabarty's teaching philosophy is simple, but it can be extended throughout his role: "Listen to the students; talk to the students—not at them." Nicole Gerardo (associate professor of biology, Emory University) says her goal is "to get students to be able to interpret data while giving them a greater appreciation of the world around." A solid teaching philosophy is one that you can use to help make decisions based on your principles.

Learning Outcomes

I think this may be a self-evident point, but over the years, it tends to get overlooked by many seasoned professors: The purpose of teaching is for our students to learn.

How is it possible to lose sight of this goal? As we become more comfortable with teaching, it gets harder to imagine the classroom experience from the perspective of students. A common trap that professors fall into is mistaking "covering" material for actual teaching. When we explain a scientific concept, our students are likely to write it down, accept this as information, and expect us to continue. It is easy to trick ourselves into thinking that just because we say something, our students will then learn it.

At the end of the semester, it really doesn't matter how much ground we cover. What matters is the material that students learn. If we cover ground but the students don't learn, then it doesn't really count.

There are pragmatic reasons for making sure that you have clear "learning outcomes" for your course. It's not just a matter of pedagogical theory. Learning outcomes are the target you are shooting for. Your course exists for specific reasons. It was proposed to a curriculum committee and was approved by the university at multiple levels, as part of a program of study. Everybody involved in your course, including faculty who teach other courses, has expectations for what students will know and be able to do at the end of the semester. To assemble a coherent and effective curriculum (as discussed in chapter 3), you need to be as unambiguous as possible about what you are expecting from students. I think a lot of instructors have a nebulous idea of the many topics they want students to learn about, rather than an explicit set of concepts that students need to understand and tasks that they are able to perform.

Another pragmatic reason to have clearly defined learning outcomes at the outset is that this makes it much easier to evaluate student learning with assignments and exams. You don't want to find yourself teaching for weeks upon weeks, and then when it's testing time, asking yourself, "What questions should I be asking, considering the ground we've already covered?" Because if you are wondering what to put on the exam, can you imagine how the students are feeling while they're studying for it?

Bloom's Taxonomy

Now that I have (hopefully) convinced you that learning outcomes are what matters at the end of the semester, this opens up a nontrivial question: What is "learning"? What does it mean to say that our students have learned something? How you answer this question can radically change what you're teaching and how you teach.

Learning is not one thing. In this book, I'm making a point to avoid name-checking educational theory, but I'll go as far as making you acquainted with Bloom's Taxonomy, which is often applied to science teaching at the university level. It's applied so much that I suspect you've already heard of it. This theory is not a divine truth, and many education researchers think it's time to move past Bloom's. However, it's a useful launching point for thinking about how and why we are teaching.

According to Bloom's Taxonomy, there are six kinds of learning. These are the abilities to (1) remember, (2) understand, (3) apply, (4) analyze, (5) evaluate, and (6) create.

I'll walk through these six categories with an example. In geology, minerals can be ranked on a scale of hardness, which on the Mohs scale ranges from 1 to 10. Let's say the Mohs scale is a required part of the curriculum that you are teaching. What do you want students to learn about it? Do you want them to just remember the name of the hardness scale? Or do you want them to understand it well enough that they can explain how it works to others? Do you want your students to be able to know how to apply the Mohs scale to test a rock sample? Do you also want them to be able to analyze the relative hardnesses of different tested rocks? Or do you want them to be able to evaluate the ways in which the Mohs scale is useful and not useful as a tool in the field of geology? Would you like your students to be able to use the Mohs scale to create a new tool, such as a rock identification guide for your campus?

Because we have advanced training in our own fields, it's easy for us to blur the different kinds of learning for concepts that we are really familiar with. For example, the ability to understand mineral hardness and the ability to apply the concept of hardness to new situations can seem like a trivial distinction, if you're an expert. But to students, who are learning this for the first time, it's quite possible to study for the purpose of understanding, but not for the purpose of application. When students are studying for an exam, you will need to decide how much they need to know beyond the fact that the Mohs scale exists. They need to know the depth and kind of learning that is expected of them about the Mohs scale.

It is important to test students at the level at which we are teaching. Some instructors think they are being rigorous by teaching students to understand a particular concept and then testing them on their ability to apply and evaluate information related to this concept. This isn't being academically challenging—it's asking students to master something that you haven't taught! This is sometimes called "teaching low and testing high," and it's unfair to our students. In chapter 6, these issues are discussed in more detail.

What Kind of Institution Is This?

Fundamentals of good teaching are universal, but the institutional context can influence approaches that you take. Depending on where you're teaching, you'll experience different working conditions, including available resources, student preparation and priorities, latitude in your teaching methods, varying degrees of student overentitlement and underentitlement, and departmental expectations for your grade distribution. I'll give a rundown of the major categories of institutions in the United States and trends that separate the teaching expectations in each type of institution. Outside the United States, there are different kinds of universities, but this information might still help you think about the characteristics of the environment you're working in. Regardless of where you are, every single institution is unique, so remember that there is broad variation in all of these categories.

Small liberal arts colleges (SLACs) focus on undergraduate education. Teaching quality is central to the identity of the institution, and this typically comes with the expectation that you will work hard to provide a high-quality experience for students that involves a high degree of personal interaction. These campuses market themselves based on a low student-to-faculty ratio and an intimate learning environment. You can probably expect a lot of traffic in office hours, and expectations to respond to student concerns promptly and personally. It's possible that multiple-choice exams are not acceptable in departmental culture, and there is both an opportunity and perhaps an expectation that you assign and provide feedback on student writing. You are likely to spend some time fielding student requests for you to reconsider the grade that you've assigned to their work, and perhaps even parental involvement. (Some suggestions for navigating such situations are in chapter 7.)

Research-intensive institutions (including prestigious private universities and highly selective state universities) are quite different from SLACs in scale, though students may still have high expectations for resources provided by the institution. In research institutions, lecture class sizes are typically much larger. You will not be expected to invest as much into each individual student in a research institution, and students won't be able to expect much of your time outside class. However, with so many students, managing the logistics of high-enrollment courses will take a bigger front-end investment. In large classes, you are likely to have teaching assistants and graders assigned to you. If your teaching assignment features a large proportion of students who are experiencing major extrinsic pressures on their grades (such as premed students), then this might be a major aspect of building your teaching relationship with your students to consider.

The continuum between research institutions and teaching-focused universities gets blurry. If an institution places a lower emphasis on the mass production of research, then the classroom experience for undergraduates becomes a higher priority. Where teaching loads for tenure-track faculty are higher, the expectations for how faculty members interact with students shift, simply because they are jugging multiple courses at the same time. This spills over into the campus culture in general, and expectations of student teaching assistants and adjunct faculty follow suit. There tends to be less oversight about how you operate your classroom, how you design your assignments and tests. While you might be interacting with fewer overentitled students, you're more likely to experience the challenge of having students who don't come forward with difficulties that they experiencing, even though it might be possible for you to help them c

Two-year institutions (community colleges) are distinct from four-year universities in that they are wholly focused on teaching students. Class sizes are probably smaller than in huge universities, and everything is built around creating an effective classroom experience. Unlike in four-year institutions, all community college instructors are hired strictly for the function of teaching and for (almost) nothing else. In community colleges, there is often a greater appreciation of the reality that adjunct instructors are busy doing other things to pay the bills when class is not in session. Full-time instructors are probably teaching five courses per semester, and quality teaching is more likely to be valued by the institution. It is not expected that community college instructors provide individual attention to students as in SLACs, but obviously they are expected to be well prepared and take an interest in the success of their students. The politics of community colleges are quite different than at four-year institutions. In general, departmental heads have more authority and latitude to control policies. While the environment can highly vary from one place to another, the entire professional focus of the people you work with is on student success, so in a two-year college you will be free of the baggage associated with the mixed teaching-and-research mission of four-year institutions.

Most of us have a narrow experience with institutions. We have received degrees from a few universities and probably have taught in fewer. Many new college instructors have reported to me that their experiences as an undergraduate student are not that helpful when it comes to understanding the challenges as an instructor. So, even if you were a student at a SLAC, that doesn't mean you're more prepared to teach in a SLAC. It is important to remember that the culture of every institution is distinct, and carrying your expectations from one institution to another can lead to mix-ups and mishaps.

The Deficit Model of Science Communication

We learn best when we are curious. It turns out that even if we think we are aware that some information is important, it's harder to learn about it unless we are interested. When an instructor says, "This is important," or "This will be on the test," this form of extrinsic motivation doesn't activate our curiosity. It doesn't help us learn.

The underlying purpose of our job is to teach students particular things that they don't know. This puts us in a tough spot, as educators, because we're aiming to teach students very specific things, even if these aren't the things that our students are curious or excited about. Once we're aware of this challenge, then we can adjust our teaching to make the best of the situation. We can work to build enthusiasm about what we're teaching, and we can teach to make our students curious about why the world is the way it is.

In the field of science communication, the "deficit model" is the idea that we need to teach people by filling in gaps in their knowledge. Unfortunately, we can't make people learn things by giving them information. Learning happens with curiosity. The Deficit Model of Science Communication is typically applied to circumstances outside the typical classroom experience, such as visits to museums, outreach activities, science video and audio programming, and popular writing. Professional science communicators know that the deficit model doesn't work. It's important to keep this in mind as we build our entire curriculum around the deficit model.

The field of science communication is very different than science teaching. We have a very specific set of information that we set out to teach, and we have the opportunity to build a substantial two-way professional relationship with our students. Just because the deficit model doesn't work in science communication, don't abandon your expected learning outcomes too quickly. Our students are in our classes (we hope) because they actually want to learn what we're teaching. We can still take some notes from the field of informal education. Foremost, part of our job as educators is to make things interesting! Presenting information as a mystery, and then providing evidence to students that they can unravel for a resolution, can generate curiosity.

People learn science when it's part of a story. So many of the things we teach can be made more compelling placed in a narrative of discovery. Scientific knowledge is the history of people who struggled with a mystery and then solved that mystery. By placing people who were challenged by unsolved problems as characters in a story, our course content can be the plot that moves the story forward. (Please be sure that your narrative includes the people who have done much of the work but have been marginalized because of their identity, and often robbed of the credit.)

The last lesson to take from the deficit model is a general philosophy of learning. It's not helpful to see our students as empty vessels, waiting to be filled with the knowledge that we are providing. Instead, we can see our students as people who already know much about the world, and we can activate their curiosity in new directions.

A Litmus Test for Our Teaching

Every time you come to a decision point, you will be equipped with a teaching philosophy, and I suggest you ask one question of yourself as a litmus test: Will this help students fulfill the learning outcomes that you've established?

Answering this question isn't always easy. You might be surprised, however, to find that this litmus test can streamline your teaching and help you and your students have a more successful course. It turns out that a lot of what happens in college science classrooms is simply a matter of tradition. Many of these traditions are about requiring students to experience adversity and jump over hurdles, and many of these traditions have no impact on learning, or have a negative impact on learning. (For example, asking introductory students to memorize all of the molecules in a complex biochemical pathway will not accomplish any long-term educational goal; it simply amounts to hazing, by requiring students to memorize something they most definitely will forget once the exam is over.) If you're expecting students to do something because, well, it's something that is typically expected of students, then you need to decide whether this will actually help your students learn what you want them to know at the end of the course.

As you're setting out on a new semester, don't forget to have fun! Because this is a choose-your-own-adventure of your own creation, make it one to enjoy! Your attitude goes a long way in influencing the disposition of your students. Classroom enthusiasm is contagious. If you're genuinely having a good time while teaching, it creates the opportunity for students to enjoy the experience as well. If you approach the course as a predestined slog, this disinvites students from taking a personal investment. We need to meet students where they are, with respect to not just their prior preparation, but also their disposition. While I don't suggest adopting a false enthusiasm, connecting with your passion for the discipline will only help you bring more travelers on your adventure.

2

The Syllabus

Why the Syllabus Matters

Professors have remarked, "I don't always get stupid questions, but when I do, the answer is on the syllabus." Students rarely read the syllabus carefully. However, the time that you invest into crafting your syllabus will be returned to you with interest. The syllabus is as much for you as it is for the students. The syllabus provides a set of policies that, if designed well, run the class efficiently. Every semester will hold some kind of surprise. If you have policies that can deal with these surprises, you'll be spared from inventing new policies on the fly. If your syllabus is designed to deal with specific problems (and problem students), then you're already ahead of the game.

Universities treat the syllabus as a legal document. Just like a constitution is a set of rules for the government, the syllabus is the set of rules for your course. You can implement almost any policy you want in your class, though obviously that means you could have the choice of adopting unwise and unfair policies. If it's in your syllabus, it's the law of the land, so long as it's not downright bizarre or against university regulations. When "academic freedom" is invoked to defend the actions of a professor, it's usually about the right of faculty to run their classes the way they see fit. For example, do you want to ban laptops? Want to deduct points if a student doesn't bring the textbook to class or if their cell phone makes noise? I think all of those are bad ideas, but if it's in the syllabus, you can do any of those things. Because university faculty are given the academic freedom to teach as they wish, you are able to run your course the way you want and assign the grades that you want. Please use this power wisely.

The syllabus is powerful. For example, when I was in college, my Introduction to Philosophy professor required all students to know the names of everyone in the class. If I got just one name wrong on the final, I'd get an F for the whole semester. I heard that it happened to someone a few years earlier. I passed the class. But if I had botched it, then that F would probably have stayed on my record. Knowing what I know now, I'm moderately confident that an appeal would have failed, for one obvious reason: it was in the syllabus. Case closed. Setting up such a strict rule is a bad idea, because fear reduces learning.

With a well-constructed syllabus, you can spend your teaching time on actual teaching and deal with problems before they become problematic. For example, let's say a student claims an alarm clock failure and misses a big midterm exam. Perhaps they had a verifiable car accident on the way to campus. There are a variety of ways to deal with this kind of situation. With a set of unambiguous policies, you already know what to do. Moreover, your students will know what will happen too (that is, if they've read the syllabus). You can't eliminate having to deal with policy-related minutiae, but advance planning can cut this to a minimum. When you're new, it's not a good idea to stray too far from standard practices. As the instructor of record, you have tremendous latitude. Once you get comfortable, you can experiment to discover what works best for you and your students.

A good syllabus has equitable policies to help all students succeed. On occasion, you might have a student who feels that they have been treated unfairly. If they file a grade appeal, the syllabus—that *you* wrote—is at the center of the proceedings. A clear-cut syllabus minimizes the odds of this happening, because student complaints tend to take root when professors levy unfair or vague policies. If you set your own reasonable standards and comply with them, you'll be just fine.

It would be nice if your syllabus set an uplifting and inspirational note for the start of the semester, but it doesn't have to. It can just be a straightforward set of policies with all of the required elements. The syllabus is there to let everybody know what to expect. For this reason, it's important to be straightforward and unambiguous, especially about common student concerns that typically crop up throughout a semester (for example, about missing an exam, or accepting late assignments, or how a grade is calculated). Your students will be grateful for the clarity.

Classroom Norms vs. Classroom Policies

The syllabus defines the playing field for the course. If you don't want your students to treat your course like a game, then don't write your syllabus like an instruction book for a board game. The tone for the semester is set by what is in the syllabus, and also by what is not in the syllabus. While a clear set of policies is important to make sure that everybody knows where they stand, a multipage laundry list of prohibitions and restrictions can set a negative tone that gets in the way of learning.

The more you try to control the behaviors of students by outlawing conduct that you feel is inappropriate (for example, no food, no cell phones or laptops, no chatting with one's neighbor, no blue pens, no hats, and so on), the harder it will be for students to feel that you respect them as they are. For every prohibition, it's helpful if you explain your reasoning to students and get them on board. If you have classroom policies that don't have buy-in, then you'll have some disgruntled students. Aside from the impact on student satisfaction, this harms the learning environment (even for students who are not disgruntled).

How do you get buy-in for a classroom behavior policy? I suggest that you discuss this policy not as a set of rules, but instead as a set of "behavioral norms." Early in the semester, before any problem behaviors have emerged, set some time aside in class to establish these norms. Introduce why respecting behavioral norms is good for everybody, and come up with a list that you can write on the board. Be sure to include norms that respect the needs of people in the classroom who might be marginalized. For example, ask students to speak loudly for those who are hard of hearing and to keep whispering conversations to a minimum outside group work because this may distract those who are sensitive to this kind of disruption. Everybody in the room is already a highly experienced student; they know which behaviors in other students are disruptive, and they don't want them either. For some topics that don't have a clear-cut solution that works in all situations, you can see which proposed solutions work better for your class. For example, you can ask students where those using laptops should sit. I bet nearly everybody would be on board with an established norm for not having distracting material on one's computer if they're sitting at the front of the room. Likewise, students can get readily annoyed by students who interrupt the class frequently with questions, and you can work with students to establish norms to make sure that no student monopolizes the classroom. In this discussion, we should not put the onus on students to come forward to argue for basic conditions that result in effective learning, as it is our job as the instructor to create that environment. If we discuss the reasons for these norms, rather than dealing with classroom management as a set of prescriptive policies, we are letting students know that we are working alongside them to coach them in their learning.

Many a professor has a thick syllabus, one that grows every semester as they have new experiences that aren't covered by the policies they've made so far. Is it wise to just keep adding new policies into the syllabus as you become more exposed to all the possible permutations of interactions with students? Let's consider an example from a colleague of mine. This professor took attendance with a sign-in sheet. One semester, a student had missed several weeks of class and said nothing to the professor about this, but they had another student sign in for them on their behalf. When the professor deducted attendance points, the student complained that this was unfair, because the student had gotten notes and still had their work being turned in. When the student elevated their complaint to the Student Affairs office, it was dismissed. A question I would like to pose for you is this: When the next semester comes along, should this professor amend their syllabus to specify that it is not allowed to mark other students as present on sign-in sheets? My answer is no—but this was a lesson I learned the hard way.

Earlier in my career, my syllabus evolved into an overly long and legalistic document. Whenever I had some kind of interaction with students involving grades for which I didn't have a policy, I created a policy and spelled it out on my syllabus. After several years, I had very specific and detailed policies for all kinds of contingencies. What if a student athlete approached me on short notice about travel for a competition on the day of an exam? What about students who were absent on the day that the grading rubric for written assignments was distributed and wrote their papers without knowing the points on the rubric? What about students who planned to request an incomplete at the end of the semester but performed most of the coursework and received an F? I had policies for all situations like these. At one point, I realized that the text in the syllabus was not helping. With so many words about uncommon circumstances, students just stopped paying attention. I still had to deal with each situation on an individual basis, and having a written policy for every rare contingency didn't help in the long run. It is important to give students clarity in policies on the syllabus, to reassure them that you won't make capricious decisions. But you do not have to revise your syllabus every time a student comes up to you with a novel circumstance, in case a student in the same scenario will approach you in the future. It's a lot easier to develop a relationship with students where they can trust that you will be fair, if you don't approach their concerns from a restrictive legalistic framework.

Highly specific policies that apply to very narrow circumstances are off-putting. I remember on a road trip, I stopped at a chain restaurant that serves pancakes. At the front of the restaurant was a homemade sign that read "Any customer who uses a laptop computer does so at their own risk and is responsible for any damage from beverages." This sign was not useful for me or any other customers. It simply was a scar from a prior negative interaction, and it just let me know as a customer that one guest had an argument and/or a lawsuit with the restaurant about a laptop that got soaked. This sign just made the restaurant look a little less than professional, and it also let me know that if a server did spill a big glass of water on my computer, the restaurant would not do anything to fix the situation. It didn't help out the customers, and I'm not sure whether it served the restaurant well by posting such a sign. It detracted from the welcoming atmosphere that most restaurants aspire to. Restaurants don't post a big sign of liability limitations for all customers upon entry, because it's off-putting. A thick wall of prohibitive and punitive policies in your syllabus can have a similar effect.

Yes, you need policies that will handle common situations, but if you build a relationship of trust, then you should be able to handle the uncommon situations readily without referring to a thick set of rules. Moreover, a straightforward and simple syllabus helps build that relationship of trust. If a student walks into a classroom and gets handed a big, thick rulebook about what they can and cannot get away with throughout the course of a semester, this is a red flag about how the professor regards the course and treats the students. Because this relationship of trust is the foundation of learning, it's a bad idea to build a wall between you and your students at the outset.

Making Your Syllabus Your Own

Syllabi tend to follow a standard format, and there are good reasons to conform to this format. Students will be able to find information promptly. In addition, syllabi are part of the regional accreditation process that requires standard documentation. If a student transfers, a clear syllabus may be needed to prevent the student from having to retake the material they've already learned in your class. So, an exceptionally weird syllabus can be a problem.

You can look at other syllabi to make sure your syllabus fits the norm. Many universities host online syllabus repositories, and you could ask other instructors in the department to share their syllabi. For practical reasons, I recommend against copying parts of other syllabi verbatim, except for required institutional boilerplate. Individual teachers cannot be replicated, and attempting to replicate someone else's approach is not likely to work well in your own classroom. Everybody needs a starting point, and if your starting point is your own, then it'll be easier for your practices to evolve to fit your needs.

Required Materials

Be sure to list required materials high up in the syllabus and what you expect students to bring to every class session. Some universities require faculty to list the price of the textbook on the syllabus. You should be aware of earlier editions and specify which editions are allowable. Because the latest edition is typically outrageously expensive, you will be doing your students a big favor if you permit the use of an earlier edition. If you use a textbook the way most instructors do, it should make little difference if some students are reading a slightly different edition.

Every university has students who are working with an extremely tight budget. You can make things easier for these students by making a copy of the text available on reserve in the library and on digital reserve. It's the job of librarians to help with this, so don't hesitate to ask at the library. The educational allowances of "fair use" in copyright law are extensive, so you may be able to post your reading material on the password-protected Learning Management System that your campus uses.

Expected Learning Outcomes

You might be required to generate your own "expected learning outcomes," which are required for institutional accreditation. If you are developing a course from scratch, you probably have free rein to create your own. Perhaps the department will just give you a list that was already developed.

In some departments that have a set curriculum associated with a professional organization (such as business, chemistry, or education), you may be required to use a particular set of expected outcomes. To phrase your expected learning outcomes in the proper lingo, your outcomes should be written along these lines:

By the end of this semester, students will be able to:
- Build a basket made of rattan or wicker that can hold five pairs of socks
- Explain how genes and environment interact to result in phenotypic expression
- Analyze the historical relevance and literary significance of eighteenth-century Finnish novels
- Balance chemical equations

This standard format involves action verbs, with students doing something that you can readily assess. A good set of expected learning outcomes encapsulates the knowledge and skills that you are expecting from your students throughout the semester. Most folks shoot for a list of three to seven expected outcomes that represent what you expect students to take away at the end of the semester and beyond.

Schedule of Classes

The daily schedule for class activities is a staple of the syllabus. If you haven't taught your own course before, you might be surprised by how hard it is to adhere to a straightforward schedule. (We'll discuss the content and structure of curriculum in the next chapter.) If you need to distribute a syllabus before assembling a detailed schedule, listing one general topic per day (or even per week) is okay. Regardless, you'll definitely want to put a disclaimer at the bottom, saying that the class schedule is subject to change. You might want to list a vague schedule anyway, unless you have reason to be confident that you will be able to stick to a more detailed schedule.

Be sure to prominently identify major events, such as big exams and due dates for major writing projects, perhaps in **bold** or ALL CAPS. When students get their syllabi, they often (or at least, they should) fish through the schedule to copy these major events into their calendar. If you are teaching with midterm exams, then scheduling three midterms can be a plus, because it gives you the latitude to drop the score of the lowest one, which has benefits for both you and your students.

Course Policies That Anticipate Student Excuses

Even with a perfectly crafted syllabus, you're still going to have students approach you with excuses on a regular basis. They'll have reasons why the assignment is late, why they missed class, or why they want to take a makeup exam at a different time. You can take one of three general approaches to student excuses. I'm not a big fan of the first two, but that won't stop me from explaining them to you.

The first approach is a zero-tolerance policy: to simply not grant exceptions and let the students suffer the consequences. If a student slept in, or their car broke down, or they were sick with norovirus on the day of the exam? Tough. They get the grade they get, even if it's a zero. This policy is not as rare as you might expect—this is how I started out when teaching. Professors that are big on teaching responsibility to students might run their classes this way. The most common argument for this policy is an analogy: "If a student doesn't go to work they won't get paid, so if they skip class why should they earn credit?"

You would think a zero-tolerance policy would make your life easier, because you won't have to deal with excuses. However, this actually invites excuses en masse, especially if you say that no exceptions are allowed. When you prohibit any kind of exception, that's when your students will crave one the most. A zero-tolerance policy generates an atmosphere of resentment. This is bad for both the learning environment and your teaching evaluations.

A second way to deal with excuses is to field them and decide whether they are legitimate. In this approach, students won't come to you with small excuses, because they will fear that you won't give them what they want. This is why some students feel compelled to invent dead grandparents. There is nothing to be gained from repeatedly putting yourself in a situation in which you need to validate the quality of student excuses. You'll just end up rewarding the ones who are better at lying to you. You didn't go to grad school to learn how to decide whether your students are lying to you.

Fortunately, there is a third way in which you don't have to uniformly reject excuses, nor put yourself in a situation in which you must evaluate them. You don't have to punish students who have valid reasons to miss a quiz, nor do you have to bend over backward to accommodate students who are ill, hungover, or in temporary police custody.

This third way—which I advise—is to have attendance and grading policies that give flexibility to everybody without having to ask for it. This way, you won't regularly have to deal with excuses, nor with students who feel they are in need of one. Students will be grateful they aren't compelled to share their personal affairs in exchange for leniency. In your eyes, there won't be any policy difference between a car breakdown, a family funeral, and an alarm that failed to go off. If a student doesn't do something, there's simply enough flexibility in your policies that they won't need to bring it to you. Nobody will be harmed if you give everybody a little slack. Odds are your students will learn more under this policy, because they won't be suffering as much anxiety over their grades.

What does a grading scheme look like that gives everybody some leeway? You can drop the lowest quiz or two. Allow late assignments with a nonsevere penalty for being late. Drop a midterm score or schedule a date for all midterm makeups at the end of the semester. There are many variations on policies that give flexibility, and how it manifests will depend on your course. This type of approach works more easily if you have more frequent assessments (which also is a good thing, as discussed in chapter 5).

Every semester, some fraction of your students will be going through personal stuff that's going to hurt their academic performance. You don't want to give slack to students who report a dead grandparent, but then fail to give slack to the student whose parent was diagnosed with cancer, whose sibling is going through a criminal trial, or who has other challenges that do not appear on the surface but clearly harm a student's performance. The students who really need the benefit of an excuse are typically the ones who are never going to be asking for it. Consider this: If you are teaching a large class, there's a high probability that one of your students was sexually assaulted during the semester in which you're teaching. Most sexual assaults are not reported, but estimates suggest that a quarter of women experience sexual assault while in college, and 10% are survivors of rape or attempted rape. Those are horrific and sobering statistics. I imagine that you want to be able to give these students some slack, and I also imagine you don't want to have to put these students in a position in which they feel they need to tell all of their professors about it. The most humane thing to do is to provide enough flexibility in your grading scheme so that anybody going through very personal challenges will still have a chance of earning a good grade.

A flexible grading policy also accommodates the students who would have lied about a dead grandparent to go skiing for the weekend. They just won't be lying to you. Of course, some students are going through hardship and others are not, and you can't remove unfairness from life. But if you give everybody the same level of reasonable accommodation, then the playing field is more even than in a rigid system in which a single missed quiz or late assignment makes a dent in the final grade. There is no real need to run your class like a punishing initiation rite.

If you aren't flexible, students will resent jumping through hoops that seem arbitrary to them. For example, I won't forget my evolutionary biology course in my junior year of college. A couple of days before a midterm, I came down with a nasty flu. I called my professor during his office hours to let him know I was really sick and asked him what I should do. He said all I needed to do was bring in a doctor's note and I could take a makeup exam. But he wouldn't accept a note from the campus medical office, because he didn't trust their judgment. So I needed a note from a doctor off campus. While a bunch of my fellow students had a doctor in the family they could just call up, I didn't have that option. Lying in bed, horribly sick, I had to either take (and probably fail) my midterm or somehow find a doctor off campus to confirm that I was, in fact, obviously sick. It would have been very easy for my professor to confirm that I was sick. But he enforced a rigid policy to make life easier for himself and really hard on me. He said he was doing it to be fair. I believe that's what he believed. I doubt he grasped the notion that a policy equally applied to everybody could be unfair.

My professor could have adopted flexible policies that would have been easy on him, but also easier on me. But hey, academic freedom. It was in the syllabus. I eventually earned a PhD in ecology and evolutionary biology, but I feel that it was more in spite of this professor than because of him. If you want to be an inspirational professor, then grading policies that don't annoy or alienate your students are a good start. Accommodating the personal lives of students is not mutually exclusive with being academically challenging.

Missed Classes and Exams

"Did I miss anything important?" might call for a smart-aleck reply, but it's a question you'll get. When a student misses class—including one with a quiz or assignment due or a major exam—the outcome needs to be in llabus. An explicit section that explains what happens when students a quiz or exam will make it easier on everybody.

our policies should be equipped for the cusps of holidays, when attendance might drop. The week of Thanksgiving, before and after spring break, and the morning after Halloween are likely to produce more no-shows. If a student tells you, "I have a plane ticket for a family wedding and it means I'll miss the second midterm," do you know how you'll handle this situation respectfully?

Dropping the lowest score is simple, but if you don't take this approach, how should you handle a missed exam? Do you offer makeups, and if so, how are they scheduled? It's easy to get yourself into an organizational mess over dealing with students who miss exams, even if you have a small class, because it's only natural for students to interview classmates who have already taken an exam. You need to find your own policy that works for you.

Even if students do not miss a quiz or exam, do you require attendance? If you do, be sure to explain in the syllabus how you keep students accountable and what happens when they miss class. If you don't require attendance, you should mention this too. It's okay to tell students that they are responsible for being informed of what happened in class by contacting other students in the class, to make it clear that you are not going to reteach a lesson. Consider requiring students to identify a partner at the start of the semester for this purpose.

Be sure that your attendance policy does not contradict institutional policies, as some universities have rules about add/drop deadlines and rescheduling final exams that you should be aware of.

Late Assignments

Students will want to turn in late work, so your policy on late work should be unambiguous. I think that instructors should accept late assignments for at least generous partial credit, perhaps on a sliding scale. Giving students full credit for late assignments could run the risk of enabling the students who fall behind schedule, but if you deduct too much, then you're making it too hard for the students who got behind (for whatever reason), and that deters actual learning. On the other hand, it's legitimate to not deduct anything for late assignments. Alice Boyle (associate professor, Kansas State University) "just ignores the lateness." One pragmatic drawback of that is an additional grading load at the end of the semester. I tend to deduct 10% per week, for five weeks, to a maximum deduction of 50%. If the cost of a late assignment is too high, students might not do the work; if it's too little, then they might put it off for too long.

What works for one course may not work for another. Leonard Finkelman (assistant professor, Linfield College) gave me some great reasons why he doesn't take late assignments: "First, I give students opportunities for revision, and the student wouldn't benefit from additional time. Second, I want to grade assignments fairly, and I can do that better by grading everything at once. Third, I scaffold assignments and so late work has the potential to bring down the student's scaffold." He also pointed out that it might not be desirable (or even possible) to grade a big pile of late assignments at the end of the semester. As in all matters, find what works best for you and your students and the specifics of the particular course you are teaching.

Assignments and Grade Distribution

What are students doing in your class to earn a grade? Grades are usually a combination of exams, quizzes, homework, reports, in-class assignments, attendance, projects, presentations, and anything else. The syllabus should contain a brief description of course assignments and also indicate a relative contribution to the final grade.

I'll describe a couple ways that you can distribute the relative credit for different assignments. In the first system, you merely allocate the contribution of each assignment to the final grade as a percentage: For example:

Homework: 15%
Quizzes: 10%
Midterm Exams: 35%
Term Paper: 20%
Final Exam: 20%

In this scheme, the number of total possible points in each category doesn't matter, because you're using the percent of points earned in each category. For example, the number of points in the "homework" category doesn't matter. The percentage score that students receive for homework will count as 15% toward their grade. When you write the final exam, it could be worth 100 points, or 85, or 224. It doesn't matter, because the final counts as 20%.

In an alternative system, you can allocate a given number of points to each category. For example:

Homework: 150 points
Quizzes: 100 points
Midterm Exams: 350 points
Term Paper: 200 points
Final Exam: 200 points
Total possible points: 1000

Using a point system commits you to ensuring that all graded material adds up to a certain point value at the end of the semester. Do you know exactly how many quiz points you want to write at the start of the semester? Do you think you might decide to shift the design of the homework assignments? Using a discrete points system gives you one more bookkeeping task. Whether you use a percentage system or a points system affects how students perceive how they accumulate their grade, even though there is no mathematical difference. I suspect students tend to get more grade-focused (rather than learning-focused) when every assignment has points to be won or lost. A points system allows professors to tack on extra credit within their accounting system, which isn't necessarily a positive.

There are many ways of divvying up points, and as you review other syllabi you might find a better approach for you.

The relative weight of each category should, theoretically, result in a greater or lesser time investment of the students. However, behavioral economists have shown us how emotions supersede rational decisions involving effort and rewards. For example, if you make the term paper worth half of the total grade, you might imagine that students will spend a huge amount of effort on this assignment. This isn't necessarily true. (To illustrate this point, note how much effort students might put into an extra-credit assignment worth a very small number of points.) To shape how much effort students put into assignments, be sure to discuss your expectations in class.

Participation

I recommend against allocating "participation" points. I realize that some people put "participation" at 5%–10% of the total grade, but there are a few good reasons to not include a participation score.

Keeping track of actual student participation requires a lot of time and effort and might not be a good use of your time. A lot of professors don't actually keep track of participation quantitatively on a daily basis, and just make a subjective call at the end of the semester. A lot of students realize this fact, and they think it's unfair. They think it's a way for professors to tip the balance of their gradebook for, or against, some students. "Participation points" are widely perceived as a "fudge factor," and that's because they are often used that way. Trust and respect are at the heart of effective teaching, and subjectively awarding participation eats away at mutual respect. This harms the learning environment and your teaching evaluations.

Points aside, participation is actually important for student learning. Students who interact during lessons learn more than those who are passive and unengaged. You might think that participation points would encourage the less engaged and more introverted students to voluntarily communicate throughout the semester, but it's rare. A vague notion of points at the end of the semester has little urgency on a day-to-day basis. Besides, if a student doesn't want to raise their hand to ask a question in front of the entire class, compelling them to do so to earn participation points probably won't help them learn. If you want to use participation points to reward attendance, then wouldn't it be more fair to simply track attendance?

In chapter 4, you'll be introduced to teaching methods that allow all students to be engaged in class. While having students raise hands one at a time engages a few students, there are many activities that allow you to engage everybody simultaneously—including the ones who would rarely ever volunteer to raise their hands. Participation is important for learning, and that's why you should bake it into class instead of just hoping they'll do it because points are involved.

Extra Credit

I recommend against using any extra credit in class, because it can be unfair to students in a variety of ways. This is addressed in more detail in chapter 5. Regardless of your choice, make it clear in the syllabus. If you do have a policy to grant extra credit (either more points, or something above the 100% possible score), then explain how all students have an equal opportunity to earn this credit.

Letter-Grade Calculation

You need to explain how scores are converted into letter grades. However you structure this distribution, it needs to be fair and transparent. Nothing should come as an unwanted surprise to your students, and they should be able to calculate their own lowest possible grade. The more clear and straightforward this part of your syllabus is, the fewer students will ask you to help them calculate their grades.

Fitting grades on a curve—so that a certain percent receive a certain grade (such as 20% getting As, 35% getting Bs)—is a surefire way to develop resentment, anxiety, and competition. On the other hand, a flat scale is straightforward and lets students know where they stand. A simple distribution is >90%: A, >80%: B, >70: C, >60: D, and <60: F. If you want to "curve" grades based on the distribution of scores but don't want to terrify your students, you could set a flat scale and explain that you may adjust these thresholds downward (always in favor of the students, never to give them a lower grade than they might anticipate based on their score). However, the practice of grading with very low scores, with the plan to adjust the distribution at the end of the semester, is a bad idea (and is discussed in chapter 6).

When I assign grades and adjust grade thresholds lower, I sort the final scores in a spreadsheet and look at the distribution. (I do this with the names of the students not visible to me, to make sure I am impartial.) It often happens that the distribution of grades has peaks and valleys. If that's the case, then I just assign the grades: A to the highest peak, B to the next highest peak, and so on. If it's messier, I use minuses and pluses to allocate grades between the sets of numbers that don't easily aggregate in a letter grade. Students can benefit from a grade higher than they expected, but don't assign a grade lower than you would calculate from their score.

You can create whatever distribution scheme you want, so long as it's clear and fair. Using the flat scale (above), here is an example: Everything over an 80 would have to be at least a B. But, depending on the shape of the grade distribution, a 78 could be lifted from a C to a C+, or even a B–. A score in the high 80s must be at least a B, but might be a B+ or even an A–. What would be your motivation for assigning grades higher than the flat scale? It should be because the way you created and graded tests and assignments, the final scores did not represent performance. Of course, it is overtly biased and unfair to slide the scale for individual students based on your subjective assessment of their performance separate from their earned scores.

It is likely that there are institution-specific policies that determine the impact that a particular grade has on a student. While it is our job to record the grades that students earn, when developing our policies it is useful to know how the institution interprets a particular grade. For example, I have worked in a department where a C– cannot count toward a student's completion of the major. Depending on where you work, a D may or may not count as units toward graduation. You need to be aware of these contexts if your grades are going to mean what you think they mean.

You might want to be aware of a whole other approach to grading using performance benchmarks, such as "standards-based grading," or "specifications grading." This works well when the expected outcomes are based on the ability to perform certain tasks. Because many students aren't used to it, I wouldn't advise diving in with this approach until you feel comfortable in the classroom and have read up beyond the scope of this book.

Phone and Laptop Use

The presence of laptops and mobile phones in the classroom—or their absence—will have an impact on how you run your classes and how students spend their time in the room. You need more than a policy—you need a strategy. Chapter 4 will help you develop yours. If you haven't figured it out yet when putting together your syllabus, then you can explain that mobile phones and laptops are permitted in class only with instructor permission, and unauthorized use of devices will result in ejection from class. Nancy Freeman (professor, El Camino College), makes it very clear on her syllabus: "During tests or quizzes, phones must be off and off the desk."

You will want to leave yourself some flexibility. You might have a student who actually has a good reason to have a phone nearby (for example, they might be a parent of a sick child). On the other hand, you might have a disruptive student whose mobile use is bugging other students. You'll want to leave yourself the option of telling the student to put it away (or perhaps even dismiss the student for the rest of the lesson) if they don't follow your instructions.

Personal Travel

A student might mention halfway through the semester, "I have this travel required of me," which will conflict with class requirements. Another version of this is when a student says that a family member has arranged travel on their behalf. It might be handy to put in your syllabus that if students have preexisting travel requirements, they need to tell you about it during the first week of class. If they already have something on the calendar, then you can choose how or whether to accommodate them. If a student books travel after they receive the class schedule, then they cannot hold a reasonable expectation of you to deal with, say, a weeklong wedding in a place far away.

This issue goes both ways. Many professional conferences are in the middle of the semester, and we might need to be away for one or more class sessions as a part of our academic duties. We also might have a workshop or a seminar or a wedding planned in advance. In my opinion, it's perfectly fine to incorporate your own travel into the syllabus, as long as you make sure that students aren't shortchanged. It's common to schedule exams for when travel is anticipated, and you can ask a colleague to proctor for you. You also could create an appropriate alternative activity for a planned absence. It would be a good idea to get an understanding from other people in the department about what the accepted norms are for planned faculty absences. In some places, nearly anything is fine as long as it's not a surprise to the students. In other places, the culture might frown upon faculty who are not physically present in class and advance consultation might be expected. Remember that students often have lives as complex as our own, and any flexibility that we would want to grant to ourselves also should be granted to them.

Audio or Video Recording in Class

Do you want a student recording of your class featured on YouTube? I guess you can't fully prevent this, but if you want the authority to request that a student take down a video, it would help if your syllabus said that recording is not permitted without your authorization. The way to prevent illicit recording is to earn the respect of your students, of course. Don't forget that some students might need to record for a disability accommodation, and if that's the case, they'll have the paperwork and should talk to you about it first.

Disruptive Students

It's in your best interest to include a statement that you may dismiss a student who is being disruptive, and call campus security or the police if you perceive a threat to the safety of anybody in class. (My university actually requires such a statement in our syllabi, to protect faculty members.) I hope that your classroom never experiences such a situation. I provide more guidance on such circumstances in chapter 7.

Contact Expectations

How do you want students to contact you? How long should they expect to wait before you reply to an email? Should students bring questions to you before or after class instead? Do you prefer them to call your office if they're not on campus during office hours? Do you require emails to be composed in standard written English, or are you okay with receiving an email that says, "how r u doc i have a question abt homework"? Let your students know what you expect of them and what they can expect from you. You don't have to write about this in a patronizing manner, but if you're going to be put out if your emails from students are not composed in standard written English, then please do the common courtesy of letting them know upfront.

Office Hours

In addition to listing the time of your office hours in your syllabus, this is an opportunity to set expectations. While we hold office hours for students to access our expertise, handle course business, and provide support, it might be helpful to provide a clearer expectation of how office hours are best used. How do you want to spend your office hours, and what uses are best for the success of all the students in your class? Some professors like to use office hours for tutoring students who come by for help, while others explicitly do not permit the use of office hours for supplemental instruction. Some professors lament that many students do not use office hours. When you schedule them and how prominently you place this information can matter. Formalizing what office hours are for—and perhaps not for—can help you out if you end up with students who want to use every office hour as a tutoring session.

Keep in mind that some students perceive that appointments are needed for office hours, so it's good to explain that drop-ins are welcome. Because office hours can never work for everybody's schedule, be sure to clarify that you are available by appointment at other times. One way to encourage students to become familiar with using office hours is to have students sign up for a time to visit your office hours in the first few weeks of class, so that they know where your office is and that it's okay to visit.

Learning Management System

If you are using the university's Learning Management System (LMS) for your course, be sure to specify specific expectations in your syllabus. Do you expect your students to log into the system once per week or once per day? Do you require that all assignments be submitted through this system? Chapter 4 has a section about the LMS.

University-Required Elements

Be sure to find out if the university requires some specific boilerplate. This could involve academic integrity, honor code, disability accommodations, campus safety, or sexual harassment. When you include it, a copy-and-paste job should be fine.

Accommodation for Disabilities

Universities typically require a statement about accommodations for students with disabilities, but even if it's not required, I imagine that you'll want to accommodate disabled students. Professors are typically required to provide accommodations prescribed by the campus disability office, such as extra time, a quiet space, or a computer. Keep in mind that the process of getting diagnosed might be onerous or expensive, and many disabled students are reluctant to be processed through university bureaucracy or are unaware of available services. Moreover, students who are anxious about serious disabilities might not feel comfortable approaching you. A couple sentences that show you are sensitive to these concerns can go a long way. For example, you could say in the syllabus, "I expect that this class will have students with disabilities that require accommodations for you to maximize your success. If you have unmet educational needs related to a disability, please let me know so that proper accommodations may be arranged. For more information about disability accommodations available to all students, students are referred to the section in the University Catalog to review the available accommodations, under the section titled Disabled Student Services. The university adheres to federal, state, and local regulations to provide reasonable accommodations for students with temporary and permanent disabilities. Formal accommodations for students with disabilities are available through the Disabled Student Services office, which is located in the administration building." In addition to putting this text in your syllabus, I think it is critical to mention your openness to accommodating students with disabilities on the first day of class. Because disabled students have likely had adverse experiences with professors who are resistant to supporting their particular needs, letting students know that you support them can make a big difference.

Activities Outside Class Hours

If your class has activities that take place outside scheduled class meetings, this needs to be disclosed upfront. Is there a lecture series that students need to attend, or a weekend field trip? Students have other courses, and commitments outside coursework. It's unfair to spring this on them after the semester has started, especially if it's connected to their grade.

Lab Policies

If you're teaching a lab section, many of the policies designed for lectures might not translate well to the lab. If multiple sections are being run, some students might assume that they can just attend a different lab section, so it's a good idea to be upfront about whether switching lab sections is permissible, and if so, under what conditions. Also, if you have any specific policies on lab attendance, late arrivals, minimum time spent in lab, and the like, be sure to spell these out clearly. Broader issues about teaching lab sections are discussed in chapter 4.

Academic Misconduct

Here is a guarantee: You can expect cheating. That's because nearly every class has cheating. I'm not kidding. Cheating is standard operating procedure in American education. This isn't my opinion; it's just a simple fact, established by the people who study this for a living. Chapter 7 discusses strategies to handle and minimize cheating. What do you need to say about cheating on your syllabus?

Your syllabus needs to state the unambiguous consequences for cheating. If you don't spell it out in the clearest and most obvious terms, then your policy might not stand up to scrutiny. Universities allow professors a huge amount of latitude in how to deal with cheating and often require you to report it (even though most faculty fail to do so). Sanctions for cheating can range from a slap on the wrist to an F for the entire semester. I recommend that you list in your syllabus the most severe consequence possible that you can administer with the course, such as an F as a final course grade. (Other, more serious consequences, though this almost never happens, would come from the authority to which you report academic dishonesty.) The pragmatic reason for listing this severe penalty is that you want to leave yourself this option. Even if you're not inclined to flunk a student for cheating, you can imagine that there might be very egregious circumstances that would merit such an action.

Plagiarism

Cheating on exams is common, but plagiarism on written assignments is even more so. Some students know it's not proper, but throughout their high school and college careers they are rarely, if ever, called out on it. Moreover, a lot of college students have never developed the writing skills to be able to write without plagiarizing. Though it might sound farfetched, a lot of students actually don't realize that a copy-and-paste job from Wikipedia is plagiarism, so long as they do a light edit on the text they've copied.

The way to prevent, and deal with, plagiarism is discussed in chapter 7. In the meantime, your syllabus needs a commonsense definition of plagiarism. You could place a signature page at the back of the syllabus that students are required to sign and return, acknowledging their awareness of the plagiarism policy and the definition of what constitutes plagiarism. You may also require students to complete an online tutorial to train them how to prevent plagiarism.

When I said that universities treat syllabi as contracts, this is particularly true when it comes to policies for plagiarism. I imagine that you don't want your plagiarizing students to claim that they were not aware that they were plagiarizing. If your syllabus has a clear description of what is not permitted, then your students cannot defensibly make such a claim. Of course, it is quite likely that students might not be scrutinizing this section of your syllabus, but having a clear statement about a plagiarism policy is important if you ever end up dealing with a case of plagiarism.

You need to specify a penalty for plagiarism. I recommend putting on your syllabus the most extreme penalty you can imagine—the same as for overt cheating—because if you don't specify the maximum consequence, you will not be able to implement it if necessary. For example, let's say you discover that a student wholly downloaded a file from a term-paper mill and only changed their name on the cover sheet. And let's say you preceded this event with a detailed in-class exercise, in which students unambiguously learned that this practice is a forbidden form of academic misconduct. I imagine you'll want to leave yourself the option of flunking this student.

Study Tips

The syllabus is a good place to put information that you are likely to find yourself repeating over and over again throughout the semester. This is why you want to put useful advice on how to study for your class. Students often perform poorly not for lack of effort but because they don't know how to study effectively. Particularly after the first exam, it's likely that students will ask, "How do you think I should study?" While you should feel free to have an extended conversation, perhaps you could start with "Have you tried out the study tips in the syllabus?"

Some students merely rewrite their notes, read the textbook over and over, and perhaps highlight their book. Some students might make flash cards or look up a bunch of videos. If there are study approaches that you want your students to use outside class, your guidance is necessary, and you can provide specific examples. For example, students can work in groups to develop problems and quiz one another. They can create concept maps connecting important terms and ideas. They can write mock test questions for one another. They also might be able to work on problem sets in the book or set up a discussion group with classmates. You can also provide links to sites that provide good study techniques.

Campus Resources

Our institutions often have resources available to support students, but they are not aware that these resources exist or don't know how to access them. A growing practice is for instructors to include a page in their syllabus to direct students to resources that they might need on campus, providing specific locations, hours, phone numbers, and email addresses. This should include food pantries, mental health and crisis resources, student resource centers (for example, centers for disabled students, women, undocumented students, LGBTQIA+), and academic support (writing center and tutoring offices). Even if students have easy access to this information elsewhere, providing this to them still might help, and it also signals to students that you care about their welfare beyond the bounds of the classroom.

When in Doubt, Include It

Is there something else that you think you should put in your syllabus? Put it in! The syllabus is there for your students so they will be able to have a clear and predictable semester and won't get blindsided by arbitrary policies. The syllabus is there for you, so that when you carry out your sensible policies, the students will (or, at least, should) know in advance the consequences of their actions or nonactions. The time you invest into constructing clear policies for your syllabus will be repaid by being able to focus on the content of the course itself, which is the topic of the next chapter.

Updating the Syllabus

Once you have finished developing your syllabus, try not to think of it as a mature document. We evolve as teachers, so our syllabi will evolve too. A complete syllabus is a template for future semesters, and changing the syllabus is the way we change the mechanics of how we run our classes. The start of a new semester is a chance to reflect on what changes you wish to make to improve your class.

When I started teaching, I adopted many policies from some of my favorite professors, which worked well for me as a student. As an instructor, I found that these policies were not necessarily the best for my own style or did not work well for all of my students. It took several years before I found an approach that worked well for me, but I still make changes as I learn new things. Each semester you will encounter a situation that might require you to amend your syllabus. For example, if your class is in a time slot that tends to conflict with activities for student athletes, you might want to develop and spell out a policy for them.

The most significant part of your syllabus that you'll want to regularly update is the lesson schedule and assignments. If you get to teach the same course on a regular basis, this will save you the trouble of reinventing a wheel that you recently built. If you keep a copy of your syllabus handy and make notes on it throughout the semester, the second time you teach a course it will be easier and better.

3

The Curriculum

How Do You Decide What Matters?

You're the expert. What happens in class is up to you. You have a limited number of hours in class for the semester. In addition, there is an expectation for students to work on the course outside class. So, given these time limitations, what do you teach?

Depending on your teaching assignments, your scope may be highly prescribed. Even if it's not required that you teach a set curriculum, it's possible that a full curriculum has just been plopped in your lap. Alternatively, you might have to develop a full course from scratch. Most of the time, we're dealing with something in the middle. For example, the textbook for the course is already picked out, or you are required to cover a number of defined topics, though exactly how you go about it might be up to you. If you're in a field that has departmental accreditation, then the content of your class is circumscribed to some extent by the accrediting body. For example, the American Chemical Society (ACS) reviews the curriculum of a chemistry department, assuring that there are common elements in all general chemistry courses in approved programs. Regardless, you've got a lot of decisions to make.

You presumably have plenty of opinions about the content that you want your students to learn. If you're like most people, the real challenge is narrowing down the material to something that can fit within the available time. As you make decisions, try to remember that you have already traveled far on the road to expertise. Experts tend to have a hard time remembering what it felt like to be a novice in their academic specialty, and this can sometimes result in classroom curricula that don't meet the students where they are. There is a huge difference between understanding something and knowing how to teach it. For example, even if you understand many sophisticated details about how electron orbitals work, that doesn't necessarily mean that you know which approaches to teaching orbitals tend to be most effective. Every discipline has a well-developed body of scholarship on effective teaching approaches for specific content. If you're looking for some of this specialized information, it's often called "pedagogical content knowledge," shortened to PCK.

Unless you're teaching a brand-new course, your course is probably a well-established cog in the machinery of the departmental and university curriculum. In theory, you can teach whatever and however you wish, because that's part of your academic freedom. In practice, there may be some very specific expectations for what you'll be teaching. This is especially true if your course is a prerequisite for other courses in the major, or part of a multicourse introductory sequence. Before you do much work to build big plans, be sure that your main objectives fit within the needs and expectations of your program.

Backward Design for Curricula

When I started out teaching and was developing new courses, I thought of all the important topics that I felt belonged in the course, and I put them into the schedule. There wasn't the space, so I dropped a few things and rearranged and squished the pieces together until they could fit into the available time. I think this is what most people do: Just come up with a big list of topics, and then trim down until it fits. I don't think this is teaching malpractice, and classes can work just fine this way. Nevertheless, teaching a class with this kind of curriculum is like going on a road trip with just a map and nothing else. You cover a lot of ground, see a lot of sights, but is there a destination? I think our classes can serve our students better if we have a strong vision from the outset.

Once you have identified the major learning outcomes for the entire course, you can chart out the steps that you need to take to arrive at this destination. This is what curriculum design folks call "backward design" or "backward planning." In the previous chapters, I've mentioned how it's important to develop expected learning outcomes. Even if you aren't expected to formally list them on your syllabus, it's solid practice to be able to say what you expect students to get out of the course at the end of the semester.

Here's how you do backward planning: Specify what you want students to get out of your course, in the big picture. Ultimately, I bet you can name a relatively limited set of concepts and skills that you definitely expect your students to get out of your course. That's your destination. Once you have that destination, you can flesh out the pieces that are essential pieces for arriving at that destination. Let me illustrate with an example.

Let's say you're creating a curriculum for a course in physical geology. And let's say you have decided that your big-picture take-home lessons at the end of the semester are that your students (1) have an appreciation of deep geologic time, (2) understand physical processes that make rocks the way they are, and (3) understand the observable mechanisms that shape the major features of the planet. That's only three things, but those are three really big things. What you need to do next is take each of those things and break them into the pieces that you will need to teach to get students to that endpoint. For example, for the processes that make rocks the way they are, you'll need to cover the kinds of rocks and their properties; the mechanisms by which rocks are formed, transformed, and destroyed; and comparisons of locations that have experienced different rock-forming processes. A geologist could create 10 hours of classroom lessons to deal with these topics in adequate detail for an introductory course. If you flesh out the other two major goals, that would probably fill out the semester quite well.

Even if you think about your semester using backward design, you probably will still have too many topics you'd like to cover, and are facing the question "What should I cut?" Backward design helps you decide how critical a topic is. Instead of just relying on your subjective opinion as an expert of what is important, you also can figure out whether or not a particular lesson will fit into a plan to create a larger understanding of a few central topics. This can be a painful process, because you might discover the content that you deeply love is not so important for fulfilling your course-level learning objectives.

Covering Material vs. Teaching Content

How much material do you plan to cover throughout the semester?

That's actually a trick question! How so? In my opinion, the answer should be none, zilch, zero, zip, nada. If we're teaching to "cover material," are we really teaching? Yes, this is semantics, whether we are "teaching" or "covering." However, I think when we say we are covering material, we are implicitly saying that we just merely talked about it.

Traditionally, this dilemma has been framed as "depth vs. breadth." In that context, the issue is how many topics you "cover," and how many details you cover within the topic. In the "depth vs. breadth" trade-off, the idea is that you have a set amount of information that you can cover, and the issue is whether you share a lot of information about many different topics, or you share an equivalent amount of information on a smaller number of topics. I think the "depth vs. breadth" trade-off can be a useful concept, but it doesn't address the issue of teaching effectiveness or how well students can learn throughout the semester.

I'm suggesting we consider a different trade-off to replace the "depth vs. breadth" trade-off: a "covering vs. teaching" dichotomy. While you will always have the same number of hours in a semester, when you "cover" material, you share a greater number of concepts and facts. When you're teaching material, you end up covering fewer nuggets of information, but the students learn them more deeply. To use an extreme example, imagine you're giving an hour-long lesson on one theory. You could lecture quickly for a whole hour, with a large number of slides, plenty of equations and principles and examples and demonstrations, and share hundreds of concepts and facts. On the other extreme, in the span of the same hour, you could give one problem to your class involving that theory and go through discussion and inquiry about this one problem in groups. Which approach do you think will result in more genuine learning?

Let's consider another scenario. Imagine your goal is to teach just one relatively narrow topic for an entire semester. Let's say you have a whole semester about the mammalian kidney, which would be an absurdly narrow topic at the undergraduate level. At the end of the semester, you should have a high level of confidence that the students in the class understand kidneys. After all, you spent a lot of time thinking, discussing, writing, drawing, and solving problems and such about kidneys. Now, imagine the opposite: Your goal is to cover animal and plant biodiversity, including the structure and function of major anatomical features. This would include the mammalian kidney, for a miniscule fraction.

If you were to compare the mammalian kidney course and the animal and plant biodiversity course, which course do you think would result in a greater sum of more genuine learning? Imagine tracking down students from both these classes, two years after they took the final exam, and evaluating how much they learned from each course. I would bet you that the students in the mammalian kidney course remember more about the kidney than the students in the animal and plant biodiversity course learned about biodiversity. That's because the biodiversity course covered a lot of ground, and it wouldn't be possible to teach everything in that class in a way that would result in deep learning. On the other hand, the kidney course would give students the chance to think hard and critically on a regular basis. Less information was covered, but it was explored in more depth. A course about all of life, at least the way I think it is often taught, is skimming the surface of everything. Some departments will call such a curriculum a "March of the Phyla." When you're marching through a lot of territory, you're not going to remember it all that well.

Covering something merely for the sake of covering it is pointless. Having a student be introduced to the concept of something, but not taking the time to actually support learning, rarely does anybody a useful service. This leaves us with a real problem. The classes we teach typically are not about something as narrow as the kidney and are often far more like the March of the Phyla. How can we possibly do justice to such a broad topic, if we don't make the point to cover a lot of ground? This, I think, is one of our central challenges in teaching. When we decide the quantity of material we are going to teach, that inevitably structures how we teach it. If you have to cover a lot of ground, then it's hard to teach in a manner that promotes genuine learning.

In curriculum, less is more. The more successful you are at paring down what you are going to be teaching, the more you can teach so that students can learn it deeply. I think it's possible to teach a wide range of ideas under a central theme but still have depth of learning. How do you do this? You have to take care in selecting a small number of central topics that feed into the central goals of the course, and teach those really well, and let a lot of the details go. It's really hard to let the details go, as experts, because we've fallen in love with those details. Our students, however, might learn a lot more if we focus on the big concepts and then use some fun details for engagement and application. Ultimately, as a general principle, it might take an hour to deeply teach one fundamental fact or concept. That would mean that, over the course of a semester, you need to pick out about 40 major facts or concepts and let everything else go.

Let me illustrate the "coverage vs. teaching" with the example of the introductory course on the March of the Phyla. The standard March of the Phyla course would involve a few lectures on the diversity of animal groups. (The class is never really titled March of the Phyla; that's just the nickname used by old-school biologists.) These lectures would involve detailed information about all of the common groups of animals, and often the subgroups within each group, and evolutionary innovations found in each group that allow us to distinguish them from one another.

I taught a March of the Phyla class for a long time, to lower-division students. I then would often see these students again in the upper division. One student was doing well in my upper-division behavioral ecology course and also received a strong A in the March of the Phyla. They asked me with full sincerity, "Is a spider an invertebrate?" I did my best to try to not look shocked or annoyed, and I asked, "Think back to March of the Phyla. Do you think you could sort this out from what we did in this class?" And they said something along the lines of "Oh, that was two whole years ago; we talked about so much in that class I don't think I could remember that kind of detail."

I don't think I can hold that student at fault. I hit them with a tsunami of information, expected them to learn it all, and after the semester was over, there was only wreckage left. To be fair to myself, this is standard practice, and it was expected of me by my department, but it still wasn't the best way to help my students learn. What could be the alternative?

How could I possibly encapsulate all of the breadth of life in just 40 fundamental facts and ideas? That sounds impossible! That's because it is. What's even more impossible is coming up with 3000 facts, cramming them into 40 hours of lecture, and expecting anybody to genuinely learn all of it. Instead of packing a mountain of information into a few lessons about the evolution of major groups of animals, I now have identified three major lessons that I want students to take home. (In case you're curious: the evolution of structural complexity in organisms allowed them to manage physical challenges in the environment; nearly all species that have existed are now extinct; most animals are arthropods.) This might sound like a narrow scope of concept for an introductory course. However, these same students, who are asked to read the textbook and told what to study, can do just as well on the same exam administered to the students who undergo the March.

This approach to curriculum is most difficult when you are expected to teach to a set curriculum that you have limited control over. If you're teaching one of many sections, this might apply to you. If you look at the statements from major professional organizations about undergraduate teaching, this isn't the message you get. For example, the ACS accreditation guidelines say, "The diversity of institutions and students requires a variety of approaches for teaching general or introductory chemistry." ACS does prescribe a number of topics for lecture and skills students must learn in lab, but it's very short and allows for a ton of flexibility. Ultimately, if you want to teach a smaller set of concepts with a goal to promote deeper learning, it's okay to experiment.

Using Learning Objectives for Each Lesson

Every time you step into the classroom, you have a specific plan for something you want students to learn. Hopefully, I just convinced you to narrow down that something, so you have the opportunity to genuinely teach that concept effectively in the span of one class session. Ideally, your learning objective for a single lesson can be distilled into one or two sentences.

Some of the most effective instructors I've seen start the beginning of every class by displaying the learning objectives in writing. They write it on the board, leaving it there for the whole class session. It will say, "By the end of this class you will be able to ____." For example, "By the end of this class, you will be able to explain how radiometric dating works," or "By the end of this class, you will be able to diagram ionic bonding and describe circumstances when it will occur," or "By the end of this class, you will be able to write the equations for exponential and logistic population growth and provide novel examples of carrying capacity in nature."

The reason these instructors put the learning objectives up for the entire class is to help students be more involved in their own learning. If students don't know where the lesson is going, then they'll have trouble following along. When students are aware of the destination, it's a lot easier for them to keep track. Students are going to be more engaged if they can see where you're going, and this engagement isn't about entertainment, it's about effective learning.

Even if you don't end up writing your learning objectives on the board for every class, it's a good idea to have a clear idea what those learning objectives are, to steer your own instruction. You don't have to obsess about every word that comes out of your mouth to make sure it articulates to your learning objectives. Asides and short intellectual cul-de-sacs definitely have their place in class. These learning objectives are necessary for you to have developed a coherent lesson, and to help you avoid overwhelming the class with a mountain of detail. We're teaching this stuff because we love the details. Perhaps some of our students will love our content as much as we do. Still, that does not give us license to riff for an hour without a specific plan.

For example, as an entomologist, I could walk into a classroom with no advance warning and wax semieloquent about many specific topics involving insects (e.g., vision, or reproduction, or population ecology) for a whole hour. That obviously would be a huge disservice to my students. This wouldn't be a lesson, it would just be a mostly unfunny insect stand-up routine. This is an extreme situation, to think that one would walk into the classroom so unprepared. On the other hand, a lot of instructors will walk into a classroom with a set of slides that outline the content they wish to teach, but without planning how they wish to teach it or how long it will take. So, then, what's a proper way to figure out how to narrow the content related to the learning objective?

Using the same example, let's say I'm teaching an actual insect biology course and have a single lesson dedicated to vision. Without thinking hard, I can think of so many consequential things I want to say about insect vision that it would fill several hours, if I discussed them in enough detail. This includes the structure of insect eyes, how insects can see polarized light, where vision is processed in the brain, and the visible wavelength spectrum, for starters. This steers me right into the danger zone, where I feel like throwing facts at students instead of building a coherent lesson. This is the time for me to temper my enthusiasm with the reality that I can only say so much about the topic. Moreover, I can only genuinely teach a fraction of what I can say. What will my learning objectives be? I have to ask myself: What do I think students really need to take home about insect vision at the end of the day, and at the end of the course? I've settled on "By the end of this class, you will be able to describe how insects see the world differently than vertebrates." This learning objective will allow me to introduce elements that I feel are essential and guide me when I decide what to trim.

Backward Design for Individual Lessons

Once you're confident that you've identified an adequately tight learning objective for a single lesson, you've got to build that lesson. You can do this using backward planning. It works for the entire semester, and you can scale it down for an individual lesson. Let's start with a description of what can happen when you fail to plan carefully.

Early in my career, a professor in my department was going to be out of town, and they asked me to teach their conservation biology class. They handed me their lecture notes, which were a short stack of handwritten papers. I flipped through and asked, "Which pages here am I supposed to cover for you?" They replied: "Oh, that's just the one lecture." I looked puzzled. They said, "I talk really fast. Just get through as much as you can and let me know where you left off." When the moment arrived, I taught as quickly as I ever have, going through the facts and concepts that were in the notes, in the sequence that was given to me. I got through a quarter of the material that was handed to me, tops. I frankly have no idea how a human being could have covered the whole batch in an hour. At the end, I'm not sure what the students got out of it.

The notes that I taught from were essentially a professor's brain spilled out onto paper. While all the topics were related to one another, if you were to ask the students what they learned, I imagine they could only say "a bunch of things related to the management of protected areas." Because that's all the lesson was, a hodgepodge. By the time the course was over, I doubt they really would have remembered anything that we covered in that class. You don't need to make each lesson a masterpiece, but it's a good idea to identify a main idea that you're planning to teach before you step into the classroom.

For each lesson, you can use backward design to make sure your teaching plans will lead your students to the main point of your lesson. This isn't more work than writing any other kind of lesson. It's just a way to organize your thoughts to make sure they connect as clearly for your novice students as they do for you. Ideally, most of what you do in the course of the lesson is contributing to students coming to a strong understanding of one central concept. Some examples might be "Plants grab carbon dioxide molecules from the air, and in the light and dark reactions, strip off the carbon and release oxygen as a waste product," or "The SN1 reaction is a nucleophilic substitution reaction, with only one of the compounds limiting the rate of the reaction." Another one that might work is "The evolutionary history of the genus *Homo* involves a number of species that emerged in Africa, with multiple dispersal events that happened within the last few million years." These are not phrased as performance-based learning objectives. Ideally, you can easily shift these statements into a "Students should be able to . . ." statement, so students will know what level of learning is expected of them.

Building off that last topic, let's say you only have one or two days to cover human evolution. There is so much to know! There are entire classes in anthropology departments on this topic, involving very thick books! Moreover, as new fossils and new research techniques emerge, we are learning new things on a regular basis. So, how can you best try to encapsulate this rich detail in such a short time span? It would be really tempting to just give students a big tour of the big moments in human evolution and a variety of cool facts and make sure that they memorize certain things for when the exam comes around. Alternatively, you could pick out pieces of information that build to the one main thing you want them to take home. This, of course, can still involve a lot of cool stories and interesting facts, but they could be ones that directly support your main point for the lesson.

How Much Memorization?

Even though there is no purpose to covering material for the sake of covering it, this doesn't mean you have to teach all concepts at a deep conceptual level. In the first chapter, I introduced Bloom's Taxonomy. For much of our content, we will want our students to simply remember and understand a lot of information. This is particularly true in lower-division courses targeting majors, because they are learning the building blocks for the higher levels of learning in future courses.

Before you ask students to memorize something, consider if this is genuinely a critical expected learning outcome for students at the end of the semester. For example, in introductory biology, it has been traditional for instructors to require students to memorize the chemical structures of the molecules involved in the process of cellular respiration. Cellular respiration is the process by which all organisms take a sugar molecule and release its energy for organisms to work. It is clearly a foundational process underlying the existence of all life. However, in the long run, does it really matter whether introductory students have memorized the chemicals involved in every step of respiration? I remember as an undergraduate, I had to memorize this and write it down when I took a test. Now, as a professional biologist, I look back at this kind of memorization as ritual hazing. It was a hoop I had to jump through that, frankly, did not provide a greater understanding of how respiration works. It was something I promptly forgot, because it didn't matter.

On the other hand, I also recall being expected to memorize the characteristics of many plant families, while I was taking a course in Plant Systematics. I was required to know the way to distinguish dozens of different kinds of plant groups, based on the arrangements and shapes of their leaves, the structures and parts of their flowers, and their overall appearance. This was a slog at the time, and to receive a decent grade in the course, it also was a hoop I had to jump through. I've forgotten some of it, though I have retained a fair bit of this information. I think this kind of memorization was completely reasonable for the course.

Why would I think that memorizing the molecular structures in respiration was not a good idea, but memorizing the traits of plant families was a good idea? It's because knowing the chemical structures involved in the reactions of respiration was not one of the major learning objectives for Introductory Biology, whereas being able to identify plant families was one of the major learning objectives for an upper-division course like Plant Systematics. Knowing the molecular structures involved in respiration would make sense for an upper-division course such as Cell Physiology.

How much memorization should students have to do, for example, in introductory physics? Though it's been a long time since I took physics, I'm not going to forget something as fundamental as $F=ma$. However, I was never required to memorize this equation. In my college physics classes, the professors operated the course so that we all could bring in a single "cheat sheet." The emphasis was on understanding the principles of physics and being able to apply them to problems. If we only needed to understand material at a basic level, then we were asked more basic questions, but we still could bring in a sheet with all of our equations and anything else we wanted to put on it. I think I learned the most about physics while deciding what to put on this paper, knowing what I would be expected to do with it! So, while I can't tell you any of the equations describing laminar flow, there's a good chance I will understand them when I look them up.

Requiring students to memorize is not inherently good or bad. The decision about whether you should require regurgitation of memorized information should be guided by the learning objectives for the course. In general, I think an introductory course might need a longer list of information to be memorized, simply because at the introductory level you are still learning the vocabulary and foundations of a discipline, and having the building blocks down pat is important to be able to process higher concepts. But memorizing details doesn't make much sense when one can look them up in just a few moments. Unless we are preparing our students to found a new civilization in a location where there is no access to the internet, the educational rationale for memorization should be higher than it used to be in the preinternet era.

Choosing a Textbook

Before you choose your textbook, it's a good idea to know how you're going to run the course. Is this the kind of class where you'll use the textbook heavily, assigning most of the textbook? Are you going to be assigning problem sets from the book? Are you going to be providing much other reading outside the book? This can help you decide what you're looking for in a textbook.

Some textbooks are designed to be a template for an entire course. For example, in the biostatistics class I taught for several years, the textbook was pretty much what I wanted to cover in the class. Each week my students read one to three chapters and did problem sets from the book. By the end of the semester, my students would have read almost the whole book. It got the job done really well. But on the other hand, in our introductory biology sequence, the textbook was a tome designed as reference material, and instructors just assigned small fractions of the book.

If you're stepping into an established course, and you do not have a compelling need to switch things up, please start out with the textbook that has been used in previous years. Switching textbooks is disruptive to students who are about to take the course, as well as those who have already taken the course and are planning to sell their textbook. Publishers change editions of textbooks very quickly. In most cases, the changes from one edition to the next are barely detectable. If you think an earlier edition of a textbook can meet the needs for your course, it would be a good idea to use the earlier edition. If you're not too picky, or are not referring to specific information in the textbook on a regular basis, you can allow the students to use one of multiple early editions. Considering how expensive new editions of textbooks are, we should have a very compelling reason to use a particular book to justify that kind of financial layout from our students.

If you do have a need to change textbooks, and if you have enough advance notice, you can readily size up the textbook options that are available to you. Every publisher will send you a preview copy at no cost, just by filling out a quick web form or giving them a call.

You might want to consider: Is a textbook necessary? If you don't have specific plans to use a textbook in a substantial manner, you might want to save your students the cash. There are also online depositories of free textbooks, and many of these textbooks are quite good. Many of these free textbooks are part of OpenStax, and there are additional "open educational resources" that you can identify for your class, too. If you are requiring a textbook because you think this will be a quality resource for students after the class is over, don't bother, because so many students resell their textbooks or acquire them as rentals. Online textbooks are typically formatted with print versions that can be created for students at cost, if you feel like students need to have an option for a paper textbook.

One important step is to make copies of your textbook available on reserve at the library. The reserve desk at the library might even help acquire additional copies of the textbook for this purpose. Be sure to notify your students that the textbook is available on reserve, and keep in mind that many students don't know what "on reserve in the library" means! If you do end up requiring a textbook that is expensive, this can be important to students who are stretching their finances.

Finding the Right Difficulty Setting

How much should you expect from your students? What does it mean if your expectations make your course "easy" or "difficult" from the perspective of the students? While some studies show that difficulty results in higher student evaluations, others say the opposite. How could this be? Teaching a difficult class can mean many different things. Some hard instructors have a grade distribution that means few students will earn As. Other hard instructors expect a large amount of work from students. Some instructors are thought to be hard because they keep students accountable for being prepared for every class. Others are thought to be hard because they provide critical feedback to students.

A lot of instructors might think they are academically challenging because they are requiring a lot of effort from students. However, requiring students to do a lot of work doesn't make your class an intellectual challenge. A literature class that requires students to read many long novels doesn't make the course more rigorous; it just requires more reading. Likewise, if you are requiring your students to memorize a bunch of equations, or biochemical pathways, or precise definitions for highly specialized terminology, that doesn't make your course rigorous, it just makes it a lot of work.

One common form of being "difficult" is to have high expectations for student performances on exams. This happens when instruction happens at one level and evaluation happens at another level. (For more context, see Bloom's Taxonomy in chapter 1.) Some instructors think "difficult" exam questions reflect intellectually challenging teaching. For example, if classwork focuses on understanding and explaining concepts, then the first time a student is expected to apply concepts to a new scenario, that isn't "difficult," it's just unfair. If you "teach high," then it's okay to "test high." Having difficult exams doesn't make your course rigorous. What makes your course rigorous is making sure your students meet high expectations throughout the semester, and not just on exams.

I think one way to make sure you have high expectations is to know, well in advance, what you plan to put on your exams. Have a template for the level of performance you expect from students to earn a final grade. Then, you can design your semester toward that endpoint, making sure everybody in the classroom is expected to engage with the material such that they'll be fully ready for an exam at the end of the semester. This is another way that backward planning in your curriculum helps you and your students.

Information Literacy

There's a popular media narrative that the current generation of college students are "digital natives" and navigate our sophisticated information landscape with ease. These accounts contradict the finding of library scholars, who have documented that incoming college students typically have difficulty parsing the distinction between original scholarly research and the secondary literature, evaluating the veracity of information, and finding the most reliable sources of knowledge. While many of our libraries are putting books and bound journals in storage, we still have librarians who are specialists in information literacy. It's their job to work with students to show them how to find and interpret information.

Your department should have at least one librarian assigned to work with you and your students. I think it's a great idea to schedule a field trip to the library for a presentation from a science librarian. While what they have to say will be valuable, physically getting your students into the library and talking to an information literacy expert will greatly increase the chance that your students will access this resource when they need it in the future. If your class is so huge that a trip to the library is impractical, you can always request them to give a presentation to your class.

This visit to the librarian isn't just something you can do with lower-division classes. Many upper-division students also are not prepared to navigate the primary literature and navigate databases. It might feel hard to sacrifice class time to give students skills that they, in theory, should already have. However, it turns out that most professors are glad to pass the buck on information literacy. This is also a good opportunity to become familiar with your librarian, which can pay off for your own work outside the classroom too.

Culturally Responsive Teaching

In and beyond the United States, our universities do not represent the diversity of our people. In most places, students who attend college are wealthier than average, and ethnic minorities are often underrepresented. Notwithstanding progress in some areas, students who are members of minoritized groups won't be off base if they remark that college is a club for white folks. Moreover, students who are first-generation college students typically arrive at college to discover that the prevailing culture doesn't encompass their own experiences. It's your job to make sure that all students in your classroom feel valued, and also that your teaching practices are accessible to all. It's important to emphasize running an accessible classroom because, regardless of your own background, because you're teaching college, that means to some extent you've cracked the code and you're part of the machine.

Often when we teach science, we teach the history of scientific ideas. Shannon DeVaney (associate professor, Los Angeles Pierce College) makes sure to incorporate examples of famous scientists: "Be sure you are mentioning women and people of color. Most biology professors mention Charles Darwin and Gregor Mendel by name in class. Are just as many mentioning Lynn Margulis and Rosalind Franklin? If you discuss Alfred Wegener, do you also discuss Marie Tharp? All students benefit from understanding that scientists who make important discoveries are not all white men."

Culturally responsive teaching doesn't mean that you need to make frequent references to current pop culture, nor do you have to avoid talking about old pop culture, such as 1990s music. It's okay for you to be you. Your students don't want you to pretend to be someone you are not. The first step in culturally responsive teaching is to avoid assumptions about common experiences. For example, if you return from Thanksgiving break boasting about a glorious feast with an extended family, keep in mind that some of your students didn't have such a Thanksgiving. I've also heard an instructor crack a spontaneous joke about men in prison, while students in the class have family members in prison. I remember as a student, one of my professors spoke negatively about their neighbors because they had a broken-down car sitting on the front lawn, while at the same moment, there was a broken-down car on the front lawn of my parents' house. I took this as an implicit signal that I didn't quite belong.

Science is an integral part of society and culture, and culturally responsive education means we can integrate the science of social inequities into our regular classroom experience. For example, when teaching about the biology of cancer and cell cultures, it's wholly appropriate to discuss the ethics of using cell lines originating from Henrietta Lacks, and an oceanography course can study the consequences of the Deepwater Horizon oil disaster. When discussing prizewinning scientific breakthroughs, we can point out the contributions of pioneers whose work was undervalued because of their gender or ethnicity. Culturally responsive educational practices help mitigate the negative effects of the predominantly white and male-centered perspectives in our fields and are necessary if we are to make science accessible to all.

Lesson Plans

When instructors prepare a lesson in advance, what does that preparation look like? When you walk into the room, what do you physically bring with you to keep on track? In practice, there are a lot of ways instructors handle this.

One common approach is to put everything in a slide show. Some instructors might stroll into class with just a USB drive or a laptop, or download the presentation from the server, and be good to go. Other folks will just have an outline they bring on paper. Another approach is to use standard worksheets to create a lesson plan for every lesson. This could be a premade template that you've found to fit your needs, or one that you've developed for your own approach to the classroom. A lesson plan is a loose script for what you expect to happen throughout your time in the classroom. A lesson plan is more than a sequence of talking points, because it plots out when you will be engaging with students and asking them to engage with one another. A lesson plan can have many forms.

I suggest that for every time you enter the classroom, you develop a lesson plan. Even if your lesson plan just consists of notes to yourself where you are projecting slides, it will make sure your preparation for the class doesn't fall apart because you don't remember what your original plan was. A lot of instructors will create a detailed PowerPoint with all of the information for students, but without providing instructions to themselves about how to run the lesson. In the midst of teaching, it's all too easy to skip through a part where you were going to ask students a question, or to get out of sequence. A lesson plan that has a realistic amount of time allocated to each piece will reduce the probability that you will fall behind or rush through material to stay on track.

Lesson planning doesn't stick you on a rigid track that you can't adapt as conditions change in the classroom. You don't have to sacrifice any flexibility. This framework just makes it easier for you to improve effectively. I find that teaching is exhausting and that the more detailed my lesson plan is, the less exhausted I am at the end of class. If I've off-loaded enough of my brainwork to the lesson plan, then I am less likely to overstrain my brain during class, which allows me to focus more on student engagement.

Teaching Evaluations

We need to talk about teaching evaluations early on. These university-administered evaluations, and how they are used by our institutions, are highly problematic. Teaching evaluations, as they are typically administered, are rife with biases. On the ends of the distribution, teaching evaluation scores may be informative, but for most instructors in the middle, they just aren't that helpful. My university aptly calls these evaluation forms "Perceived Teaching Effectiveness" forms. Note the word "perceived." Actual effectiveness is moot. Student evaluations of teaching are not a measure of teaching quality; they merely indicate how students felt about the course. Even in that respect, these evaluations are overloaded with biases, and universities definitely should not look at evaluation scores as a measure of teaching excellence. That said, unless you're in one of the few universities that doesn't care about these scores, it's unwise to wholly ignore them, because they probably have some consequences for you. Of course, it's not good practice to engineer our courses for the purpose of getting the highest possible teaching evaluations. It is good teaching practice to run our courses in a way that our students feel a sense of accomplishment and are intellectually challenged. That doesn't hurt the evaluations, at least, and it might help.

It might seem odd for me to put teaching evaluations in the chapter about the curriculum, considering the forms aren't that helpful. However, you do need to think about them from the outset, because, like it or not, they matter. In many settings, the only way your peers will end up evaluating your teaching performance is by what students say about you on these forms. This is a common currency of "performance" in higher education, and it's not in your best interest to ignore them.

It is strategic and self-serving to plan a course merely to avoid receiving bad student evaluations, but planning for solid evaluations can be consistent with planning to teach effectively. It is a happy coincidence that if you do the work to teach effectively, you'll also be doing things that will typically help you garner strong evaluations. If you're just planning on teaching well and having the chips fall as they may when evaluation time shows up, then that's great for you. Teaching evaluations hopefully will matter to you to the extent that you can use them to gain information to help improve your teaching. Meanwhile, many instructors can be impacted by negative teaching evaluations, and it's wise to plan ahead. As students, we aren't equipped to know how much we have learned, but we are good at gauging our own satisfaction. This is vaguely connected to learning, though, because students who are not satisfied with a class are not engaged, and students who are not engaged are not likely to learn.

The biggest way to influence your teaching evaluations is to respect your students. Unless you're a sociopath, you can't fake genuine concern and kindness for a whole semester. There is a lot that you can't control, but if your students feel that you genuinely respect them, it's highly unlikely that you'll get savaged in the evaluations, even if you had a rough semester. Students often give poor scores to faculty members whom they dislike as human beings, and that happens when they feel disrespected. The Respect Principle is about effective teaching, but a consequential side effect is also a positive impact on teaching evaluations.

If you have already taught a particular course, then feedback at the end of the previous semester can help you modify your course for a new round. But even if you haven't received a batch of evaluations, you can still build your course to plan for a successful semester. Perhaps you could think of this as another form of backward planning. Do you know what questions are being asked on the evaluation? Is your course designed so that students will think you did a good job in all of those categories?

Teaching evaluations tend to have similar questions. At my current university, the big question that folks care the most about is the last one: "How effective was the instructor of this course?" Before that one, there are a bunch of questions about how well we understand the content, if we are organized, if we were responsive to student concerns, if we were available when we said we were, and so on. A good mechanism to see if your course is ready for the semester is to take a look at your plan for the semester, and then the teaching evaluation form, and ask yourself, "Are you prepared to meet these criteria?"

It's a good idea to know in advance how the evaluations are going to be administered. There are three possibilities: on paper during a class session, mandatory online, or optional online. The way universities enforce mandatory evaluations is that they do not allow students to access their final grade report until they have completed the evaluation.

If you are in charge of administering paper evaluations in class, it would help to give some thought to timing. Are you going to do this at the start or the end of the period? Is it before or after a difficult exam or turning in a time-intensive assignment? This seems obvious, but still worth repeating: It's not in your interest to administer evaluations at a moment when students may be more stressed out than normal. A good moment might be when you hand out materials that provide information to prepare for a final exam, such as a comprehensive review sheet. For what it's worth, one study has shown that bringing cookies to class on the day of evaluations increased scores. Of course this is pandering, unless you're in the habit of bringing in treats on a regular basis. With my teaching style, I suspect that my students would find treats on evaluation day to be at least somewhat cloying, and they wouldn't appreciate a cheap bribe attempt. Once in a long while, I may bring in donuts or something else like that, but never on evaluation day. However, your mileage may vary. Hopefully you are in an environment where you don't need to worry about finessing the evaluations this way.

If evaluations are being administered online, hopefully your university will inform you when these become available to students, so you can mention this to students in class. If evaluations are not required, then you may have the problem of low sample size and a nonrandom sample. One strategy to get more students to complete the online evaluation is to provide students with enough time in class with the purpose of completing the evaluation, just as you would for students doing paper evaluations. I have heard of some professors offering extra credit to complete evaluation forms. This is presumably against the rules, and even if it's not, it's unwise because this could be perceived (and rightfully so) as a tit-for-tat gesture that might compromise the integrity of the process, even if the process doesn't have a lot of integrity from the outset.

To avoid getting a surprise in your evaluations, I suggest administering midterm evaluations, for your own use. This gives you early evidence about student perceptions while you still have the opportunity to change the course, if necessary. It helps to be open and transparent about the changes you made and why you made them. You could use the same form for your midsemester evaluation that your university uses. In my experience, when a course is on track, then the final scores are similar to what I got at the halfway mark.

I often use a supplemental evaluation form at the end of the term, which I administer to accompany the official university evaluation. I do this to separate out the competing functions of the evaluation. On one hand, the evaluation is supposed to give you feedback to help improve the course, but on the other hand, it's a summative assessment for the institution to evaluate your performance. What students might think is constructive feedback might be seen as a negative critique by those not in the classroom. Many students are not aware of how evaluations get used. It's in our interest to separate those two functions. We will get higher-quality feedback, and the university will not get mixed messages in the evaluation form.

Before we went digital, I used to hold up the university form and say, "This form [holding up the Scantron] is being used by the school as a referendum on my continued employment. I won't be able to access these forms until after the next semester already starts, so they won't help me out as I make adjustments at the end of the semester." Then I would hold up another sheet of paper [an evaluation I wrote with specific questions about the course] and say, "This one is for constructive feedback about what you liked and didn't like about the course. If you have criticisms of the course that you want me to see but don't think that my bosses need to see them, then this is the place to do it. Note that this form has specific questions about our readings, homework, tests, and lessons. I'm just collecting these for myself, and I'd prefer if you don't put your names on them." I find that students are far more likely to evaluate my teaching in appropriately broad strokes in the university form when I use this approach, and there are fewer little nitpicky negative comments that are useful to me as an instructor but irrelevant to the people charged with deciding whether I'm a good teacher. This supplemental form approach is not designed to prevent students from being honest with the university—it just gives them a chance to give me more detailed feedback and also lets them focus on giving the university what they want—their overall opinion of my performance.

In general, students give high ratings to instructors who they see as likable and approachable. There are many ways to be liked by your students, as a human being, and I think being liked is a prerequisite to really good scores. The impact of likability may explain much of the documented biases in evaluations. What it means to be professional and "approachable" for a younger-appearing female professor might look really different than for an older guy. As Susan Letcher (professor, College of the Atlantic) remarked, "As a woman, I have found that I need to dress at least somewhat formally to be taken seriously in the classroom. The men I've talked with about this issue feel less pressure to dress for the role." My experience is consistent with Dr. Letcher's. If someone had discovered a formula to remove these biases, we would already know about them. There isn't a fair solution to the problem of reckoning with inequitable norms of peers and students, and this is a problem to resolve in the context of your particular institution.

4

Teaching Methods

Teach the Way You Want to Teach

There are as many pathways to excellent teaching as there are excellent teachers. Nobody has ever planted a flag on the pinnacle of teaching supremacy. Anybody who claims that a single approach to pedagogy is the one "best practice" is promoting either their product or their brand. Instructors are different and every course has different content, and what works great for some people won't work well for others. To teach well, you have to teach the way you want to teach. You have to believe in what you are doing.

If you wanted to bake an angel food cake, you could track down a recipe in a book written by a master baker. Back in your own kitchen, if the recipe was written with enough detail and you followed the recipe to high fidelity, then you should expect to produce a delicious cake.

You may be dismayed that this book isn't just a recipe book for wonderful teaching. I do wish such a thing could exist, but teaching is not like baking. There is no single formula that works for everybody. We all come to the classroom with different ingredients, tools, and nutritional requirements. Trying to use someone else's teaching formula in your own classroom might be like trying to bake an angel food cake with corn flour and without baking soda. Maybe you could make tamales or cornbread instead.

Plenty of research has shown that some approaches are more effective than others. I've heard many experienced instructors claim that their teaching is highly impactful because they are masterful at the methods they have chosen. For example, I've heard people say things like "New approaches to teaching can work for those who are using them, but I've been lecturing for years and I know how to make my students learn just as well in a lecture format." However, if a chosen approach is inherently less impactful than other methods, then these instructors are merely making the best out of a bad choice. They are well positioned to improve their craft by learning new techniques. While some folks would like to argue that every style of teaching can be highly effective, depending on who is running the classroom, that's simply not supported by the evidence. There are some choices that don't work well for anybody. Some traditional choices are not necessarily horrible but are clearly suboptimal.

While everybody teaches their own way, there is a remarkable level of consistency in how science is taught at the college level. Our classes, for the most part, are structured like they were 50 years ago, with lectures and exams and labs and written assignments. There are many ways that students can learn, but many classes are just slight variants of the same formula, and we just shift the parameters one way or the other. There has been recent progress in the implementation of innovations that are known to work, but in higher education, change comes slowly.

I'm not asking you to break away from the base recipe and do something entirely radical when you teach. But if you look at the evidence and see ideas that have a good chance at improving on what you're already doing, I encourage you to give it a try. Ultimately, you'll be teaching your best when you believe in what you're doing.

Classroom Management

Command of your classroom is essential for learning. In education lingo, good "classroom management" means that you have the classroom under control, with students that are focused, without disruptive behaviors interfering with the learning environment. The bigger your class is, the harder it is to rein in potentially disruptive behaviors. A well-managed classroom isn't necessarily a quiet classroom. A loud room might just mean that everybody is engaged! However, since you're the one coordinating the learning environment, you need to be able to focus attention promptly.

I've had plenty of opportunities to watch other instructors teach. I found that there was tremendous variability in how much control the instructors had over the room. With some professors, when they were talking, the room was so quiet you wouldn't even realize that the hall was full with students. Then, when the students were asked to talk with their neighbors as part of the lesson, there would be loud conversation as prescribed, but when attention was directed back to the front, people would promptly switch to paying attention. At the other extreme, with other instructors, the room had a constant low murmur of conversation. Closer to the back of the room, students would be involved in full conversations unrelated to the course, and others would be watching sitcoms or music videos. Most lessons fit somewhere in the middle of these two extremes.

We owe it our students to run a tight ship. It's fair to say that the classroom with the engaged students had more learning than the classroom with the unengaged students. In poorly managed classrooms, even students in the front won't learn as well, because the instructor is spending more time and energy trying to earn the focus of the students and has less capacity to focus on the content. When I'm teaching, while I can get annoyed when distracted students disrupt the class, the ones who get really annoyed are fellow students. When someone is Facebooking in the front row, or is monopolizing discussion, the rest of the class is usually super pleased that I shut it down, as long as I do it gently and with respect. This is very difficult to finesse if you're not running a well-managed classroom. However, in a well-run classroom, there's rarely any need to contain bedlam.

Even if you've taken care to follow the Respect Principle, that isn't necessarily enough to guarantee that you will have a well-managed classroom. You need to have classroom management skills. Classroom management is a fine art that we are rarely taught. Like other forms of art, you get better at it with practice and experience. Here are some tips to help you manage your classroom. To set yourself up well in the long term, prepare well for the first few days, which sets the tone for the entire semester. Once a pattern of behavior is established in a classroom, that will steer the expectations of students for the rest of the semester. While you can lay down hard rules about behavior, it's more constructive to bring students on board by developing guidelines with your students on the first day of class. This way, they may be more invested. Ultimately, if you're going to be doing a full lecture, without any kind of break, then some students will lose their focus. By interjecting times when students are not expected to be passive, they'll have less opportunity to lose focus.

If you are experiencing classroom disruptions from students who are not focused on learning, here are some short-term tactics. If someone is distracting the class, engage with them in the content of the lesson. Overtly singling out one student is inadvisable, but changing up your lesson to make sure that this person is engaged can keep them from distracting others. For example, you can introduce an activity that requires all students to be engaged. Also, if you're standing at the front of the room and distractions are coming from farther away, you can navigate the space to get closer to the source of the disruption. Please note that a sarcastic put-down will only shame the student but not address the underlying cause. If you want to address it directly with this student, talking to them privately is the way to go.

Scientific Teaching

We are scientists. We should teach like scientists. What does that look like?

We solve problems. To do this, we evaluate what is already known about the world, and then we combine observations, experiments, and models to create new knowledge. We iteratively assess what we know and what we don't know, each time creating plans of action to take our understanding one step further. We can apply all of these principles to what we do in the classroom. It might seem self-evident that our teaching will benefit if we think about empirical evidence for teaching the same way we think about our scientific research. However, I don't think this is the standard modus operandi for most science teachers.

About two decades ago, the phrase "scientific teaching" entered the lexicon. Nowadays, you'll be more likely to hear people talking about "evidence-based teaching practices." That has a nice ring to it, doesn't it? That we might teach using evidence about what works? Now that you hear it put that way, why would anybody choose to teach without evidence-based pedagogy?

You're teaching in a science department because you're an expert in science. Likewise, there are education departments that are filled with experts in education and pedagogy. There are many journals filled with rigorous peer-reviewed research about what works in science classrooms and what faculty can do to teach more effectively. We can't accept what they are telling us with blind faith, just as we can't with science papers that we read. However, it's sensible to respect the expertise of our colleagues and recognize that their research is valid, and to develop a practical understanding of how much their findings can be generalized. The literature on science teaching at the college level is deep and rich.

If you are adopting evidence-based teaching practices, another important source of evidence is your own students. The "outcomes assessment" movement in higher education is often viewed by science faculty as a mountain of useless bureaucracy. I concede that a lot of what happens, at least in my experience, involves useless paperwork and perfunctory checking of boxes. The idea behind the bureaucracy, nonetheless, is full of merit: We evaluate the impact of our teaching on our students, and then we adjust what we do to improve how students learn! This is something that we do every time we reteach a class that we've already taught. The culture of assessment simply seeks to formalize this iterative improvement of our teaching. Anyhow, the best evidence that our teaching is effective is that our students learn what we are setting out to teach, and it makes sense to generate this kind of evidence in a way that we can improve our craft.

Using Class Time for Student Learning

In the midst of the semester, I think every instructor is all too familiar with the feeling "Oh my gosh, it is already halfway through the semester?" We only get so much time with our students, so we have to make the most of it. Once you factor in time spent on logistics, quizzes, exams, and unplanned interruptions, you will probably have between 30 and 40 hours of instructional time in a typical semester. Depending on your perspective, that could seem like a huge amount of time that you get to dedicate to one topic. I think for most of us, however, it resembles a momentary flash, that cannot possibly be adequate for the amount of mastery that we would like to develop in our students. It is important to challenge our students, but if our expectations are unreasonable, then we are not placing a realistic challenge for them. It is reasonable that earnestly and skillfully working students should be able to meet the benchmarks that we set.

Efficient Teaching isn't just about using your preparation time well, it's about using your limited time with the students well. We only have so many hours when we and our students are physically together. It's best to provide educational experiences that students are not able to easily access outside a classroom setting, with many learners together with an expert facilitator.

I recommend spending as much classroom time as possible on the process of learning. This might come off like a self-evident platitude, but I would like to point out that in the traditional college science classroom, almost no time is spent on learning. Instead, traditional teaching involves merely the transmission of information. Much of this chapter is about how we can use classroom time for learning.

Lecturing and Active Learning

To lecture or not to lecture—that is *not* the question! Much of the contemporary dialogue about college science teaching pits lecturing against active learning as competing practices. This is a false dichotomy. We all are lecturing, and we all are using active learning strategies. In every class.

You know what lectures are. Presumably, many of your undergraduate "lecture courses" were mostly that—lectures. The professor stands at the front of the room, and talks and shows a PowerPoint, and the students take notes. What makes a lecture a lecture is that the professor is talking, and the students are mostly listening and processing information on their own. In recent years, the lecture has been under siege as an outmoded and ineffective approach to teaching. Regardless, it is still the prevailing mode of instruction in college science courses. This is because, despite a lot of excitement about innovations in pedagogy, some experienced instructors are reluctant to add new elements to their lessons, or to reduce the time spent lecturing.

The main alternative to lecturing can be broadly termed "active learning." This is a coarse term that has been in use for a long time. It doesn't have a narrowly defined meaning, it's just a handy label for a broad approach to classroom teaching. There are myriad approaches to "active learning" that have been given other names. All of these terms have different connotations or interpretations that come with a lot of baggage. To some, "flipping a course" means having a full set of video lessons outside class time, and having highly structured group activities in class. To others, "flipping" just means doing a lot of active learning during class time. Likewise, to some, "team-based learning" and "problem-based learning" are technical terms referring to a very specific approach, and to others, it just means that students are working in teams, and/or working together to solve problems.

For our purposes here, "active learning" is pretty much everything that is not "lecturing." It's the least bad choice for jargon because it emphasizes that students can be observed actively doing something other than dutifully digesting a lecture. I hope we can toss out of the window the concept that there is a conflict between lecturing and active learning. Both teaching approaches coexist in all classrooms!

The comparison between lecturing and active learning has been characterized as running the classroom as "the sage on the stage" versus "the guide on the side." While these are sometimes presented as competing or conflicting approaches, all class sessions will have elements of both lecturing and active learning. For example, if a professor pauses to ask a question to the audience, and students raise hands to answer, this is a form of active learning. Whenever an instructor spends a few minutes explaining something, that's a form of lecturing.

Each time you walk into the room, you're expected to teach a bunch of stuff. Ideally, your bag of tricks will include a broad range of elements from lecturing and active learning approaches. However, your experience up until this moment might not have involved a lot of active learning. You've seen every possible way that a college professor can lecture, but you might not be as familiar with active learning techniques. This book isn't here to make you a master at active learning—though I have suggestions at the end of the book to help you out. Instead, I'll try to contextualize the trade-offs between lecturing and active learning, giving you a nudge to experiment with active learning to see how it works for you.

Does active learning fit within your teaching philosophy? Is the instructor responsible for presenting a tidy explanation for everything and the best way to learn and understand everything, or is it the role of the student to struggle with the information to understand it independently? Are you there to point students toward the pathway to understanding, or should you be escorting them all the way through to the destination? If the curriculum is a big map, should the professor lay out the map and explain it, or should the professor guide the student as they draw their own map? These questions might help you sort out how you feel about developing your own balance of lecturing and active learning.

If you have a ton of information that you plan to convey to students in class, then this is less suited to active learning. On the other hand, if you have a relatively small set of concepts that you want students to understand more deeply, then active learning might be more suited to this approach. (This trade-off was discussed in chapter 2.) There are probably some subjects where you want to introduce a lot of material yourself in the classroom, while there are some you will want to be sure that you teach very well by requiring students to think critically and deeply—this might call for an active learning component.

What is the approach to teaching that you think makes you most comfortable? What works for you, what works for them, how does it fit the course material, and how are you prepared? While lecturing sounds nice, preparing a one-hour talk for every class section is very demanding and requires a huge performance on your part. On the other hand, it might be equally difficult to develop active learning activities and manage a classroom full of interactive students while keeping the lesson on track. You're probably more familiar with the lecturing format, so it feels more natural to develop an hour's worth of a PowerPoint lecture for your class than to develop an hour's worth of activities to guide students to learn about the same topic.

Why Active Learning Elements Can Help Your Students

I think it's important that you teach the way that you want to teach. Nevertheless, I think I would be shortchanging you and your students if I didn't work to make a strong case for you to build active learning elements into your routine. That's because the educational research is piling up to suggest that including active learning into lessons is very beneficial.

Why is active learning effective? In short, students are more likely to be engaged in, understand, and retain information through inquiry-based approaches. People are a lot more likely to remember experiences, and that includes when they discover things on their own. They don't remember things when they are told about them.

In lectures—even really interesting ones!—attention can wane. It's no mere accident that TED Talks, which are carefully designed for engagement, are 18 minutes long. This doesn't mean that you can't learn from a lecture, but the attention of the audience needs to be reactivated. When instructors intersperse their lectures with clicker questions, or think-pair-share activities, or short group discussions, this can reengage students in listening.

If you're not on board with the idea that the evidence shows that an active-learning-rich course has greater educational impact than a lecture-intensive course, please flip back to the notes at the end. You'll see a reference to a big meta-analysis published in the *Proceedings of the National Academy of Sciences*. I'm not asking you to take my word for it. Please give the paper a read!

Is it possible that people can learn from lectures? Yes. Is it possible that active learning can be done badly? Yes. But, on average, is active learning more effective than lecturing? Definitely. I don't think anybody should adopt teaching methods that they think won't be effective, and I don't want anybody to teach in a way that makes them feel miserable. But I am hoping that, if you're reluctant to give some active learning techniques a go, the mounting evidence can give you a nudge.

Once I saw the results of active learning on student performance, which have large effect sizes, it became hard for me to choose against active learning. When I first started giving students active learning activities in class, it felt awkward because I hadn't experienced much of this as a student, but in time, I've seen that it works well for me and for them. I still go into lecture mode regularly, but if I find myself talking continuously for more than 15 minutes, I realize I should probably pause and give a quick problem to students to solve with their neighbors, just to make sure all of the students are checked in.

One consequence of including more active learning is that you cover less ground. This might seem like a drawback, but I think it's a reward in disguise. When we zip through a large amount of content, students do not learn much of this information—they may only retain it long enough to get through the final exam, at best. And then, poof! Gone! It's as if we didn't really teach it. So, if you can identify the key pieces of information that you want students to actually learn and take with them into future semesters and beyond graduation, they are more likely to deeply learn this information. The more active learning you use, the more you force yourself to carefully select what you are teaching and teach it deeply.

Perhaps the most critical reason to include active learning in science classrooms is that it minimizes gaps in student performance among socioeconomic and ethnic groups. In an active learning setting, students who are first-generation college students demonstrate even greater increases in learning. How we choose to teach is not just an abstract expression of our values, it has real consequences for the students in our courses. Instructors who choose to operate a lecture-intensive course are choosing a pedagogical approach that is known to increase disparities in educational attainment. Some folks have argued that it is downright unethical to teach a traditional lecture course, once they are aware that this kind of teaching exacerbates educational inequities and contributes to the marginalization of students who are in the minority on our campuses. This might be perceived as an extreme charge, but taking the evidence into account, I don't find grounds to disagree with the premise. If you are seeking a way to make your university more inclusive and to promote the success of underrepresented students, then making sure that your teaching provides them with the same opportunity to learn as other students is an impactful place to start.

Getting Active Learning to Work

Let's say you're interested in teaching with more active learning but you're not really sure how to go about it. Common wisdom is that it's easier to teach with lectures than to come up with active learning lessons. If you already have a full set of lectures in the can, then you could theoretically just recycle your old lessons. But if you are working on a new lesson from scratch, I don't think active learning necessarily has to take more time. The biggest hurdles are figuring out what to not teach and letting go of lecturing on the things that you enjoy talking about. For example, a standard class in ecology would feature a detailed explanation of a classic experiment to study competition, which involved barnacles on the seashore. Teaching the story of this experiment would take a 20-minute chunk of lecture time. The reason that lecturers would spend this time on the topic is not because this particular experiment was so important, but because the results so clearly demonstrate the principles of competition theory. If an active learning element about competition replaces this lecture piece, then why do they need to know about the barnacles? If you think that piece of content is particularly important, you can assign that section of the textbook to students, and they will be better equipped to understand it in light of the active learning lesson in class. Better yet, if you think the barnacles are so important, then you can introduce the barnacle experiment in the context of an active learning element in class, in which students have the opportunity to consider the experimental design and make predictions about outcomes. This approach does not have to be more work than preparing a quality non-active-learning lesson about the same material.

Let me give you an example about how I might teach a full lesson using a lecture and then compare that to the approach that I am using with active learning. In my biostatistics course, I've taught a one-hour lecture on the central limit theorem. (The central limit theorem explains how increasing sample size results in a more accurate estimate of the mean and also a reduction in the standard deviation.) I would provide a range of verbal explanations and come up with some PowerPoint slides and some examples, and also project some computer simulations, with virtual dice rolls, showing how the sample size affects the shape of a frequency distribution.

I've also taught the same topic using an active learning approach. Before telling students what the central limit theorem is, I'd ask students to create frequency distributions by rolling dice. (Either I'd bring a bunch of dice to class, or I'd ask students to roll dice virtually using their phones.) I ask students to quickly draw a distribution of expected frequencies if they roll one die at a time—the probabilities of getting any number 1–6 are equal to one another. I then ask them to calculate the probability of the outcomes by rolling two dice and adding the numbers together, with the probability of getting values between 2 and 12. I don't ask students to raise their hands. Instead, I ask them to work together for a couple minutes and then ask some groups to share out. I summarize this quickly on the board.

I then ask the groups to come up with their predictions about how the shape of the frequency distribution might change based on the number of dice they roll. What would it look like after just a few rolls? How about 15? How about 30? How about 100? Unless some students have read further ahead in the book than I expect, there ends up being a broad range of predictions, and in my experience, few of those predictions are actually correct. Then I ask students to build their own frequency histograms by rolling many, many dice. I get some of the student groups to draw their plots on the board when they are done. That dice rolling and data management takes at least half an hour. (Alternatively, there are websites that can be used to generate these frequency distributions in just a second, using random number generators. That's also a valid choice.) Then I ask students to compare their findings to their predictions. Once we put everybody's findings together, and groups mill around to visit with neighboring groups, then people figure out how the distribution changes as they roll more dice. Then, as the class is winding down, I go to a website that does virtual dice rolls to cement the discovery, and we spend a few minutes talking about how the central limit theorem seems counterintuitive, but indeed that's the math and the probabilities work.

There are some big differences between the lecture and the active learning lesson on the central limit theorem. In the lecture, the students were exposed to a variety of examples, they saw a greater variety of computer simulations, and they had me explain it to them in a variety of ways. With the active learning approach, I didn't have the time to show them the central limit theorem in its full glory, but I did give them the opportunity to think about it, struggle with evidence, and figure out what is going on independently. I can rest assured that if any student wants a lecture, they can always hit up the internet, and I also assign the textbook to cement and supplement the active learning lesson. But if they are to benefit from an activity to interact with peers to discover information in a structured manner, then this is something that is best facilitated in the classroom.

How can we teach other kinds of topics using group work and active learning? I offer more suggestions in this chapter, though effective active learning approaches might differ substantially depending on the nature of your content. Developing these lesson elements may be more efficient than lecturing. I've found that, at least in my own experience, building a robust active learning lesson doesn't take any more effort than building a high-quality lecture for the same time period. Sometimes, it might even take less work! Of course, once you have a lecture course well prepared, it will take much supplemental effort to add new active learning elements. This might be incentive to build active learning into new courses from the beginning.

When you use active learning elements as a part of a lesson, you're just increasing the size of your teaching repertoire. Here's an exercise-related metaphor: If you go to the gym and you are told you need to exercise more, then a trainer might prescribe a regimen of cardio machines, weight lifting, and swimming. Which might not sound fun if you're doing it on a regular basis without any variety. But here's the thing: You don't need cardio machines and a swimming pool. You just need to exercise. You can go for a hike, run, walk, take a bike ride, go play racquetball, or something else. Whatever works for the particular moment. I think active learning has the same perception issue: You don't need to follow anybody's particular script. The edupreneurs sell their brands (such as "Peer Instruction" or "Learning Catalytics") as rigid approaches, but really, it's as much of an art as a science. You just need to engage students to interact with the material. Just like you don't need to buy a fancy home gym to get a good workout.

Keep in mind that active learning can be exhausting for students. One of the main reasons to use active learning is to make sure that students do cognitive work, and it's hard work to be on for a whole hour at a time. If your class has a bunch of active learning in it, then you can expect that your students will emerge more exhausted than if you just talked for an hour. This isn't necessarily a lighter cognitive load for you as a faculty member, because even though active learning means that you're spending less time performing, managing active learning activities can still be taxing.

Your Bag of Tricks

Running a classroom is a kind of performance. Like all performers, we have a repertoire that we rely on, just like dancers have a number of moves they use. In teaching lingo, this sometimes is called a "bag of tricks." It's nothing magical or tricky; it's just the full range of activities that you do with your courses.

Of course, you're already familiar with the broad repertoire available to you. After all, you've made it all the way through college; think about how many professors you've seen teach! Be careful, though—novice teachers often make the mistake of thinking that a well-selected bag of tricks constitutes a coherent teaching approach. Just like you wouldn't write a song using the sounds assembled from your five favorite songs, your teaching performance can't be made up of spare parts from your bag of tricks.

Your choices are best grounded by your overall goals for the course. This is where a teaching philosophy can help out. Of course, when you hear of a good trick, feel free to use it! We get better at anything by trying and failing—and the same is true for teaching. So, feel free to try a new trick, and if it doesn't work out, then improve on it or drop it.

You can learn about new classroom interactions by talking to other faculty and by checking out the suggested resources at the end of this book. Some are very quick interactive bits that only take a minute or two. For example, a "muddiest point" can happen with a few spare minutes at the end of a lesson, when students write down on a sheet of paper the part of the lesson that they understood the least. You can go over these the following period.

Whenever you have an impulse to ask a question of your students, you can do a "think-pair-share," by letting each person have a short moment to think about it and share with their neighbor; you can then choose pairs to report back to the group at large. Think-pair-share is very common in lecture-based lessons that are interspersed with activities.

For more substantial problems where the consensus is unclear, an additional twist on think-pair-share is to pair up groups of two into groups of four and to ask the groups of four to come to a consensus. This takes more time than just telling students what the correct answer is, but if they struggle through the ideas by discussing them with others, that's what learning looks like.

You can ask your students to look up information on their phones or laptops. A wide range of old-school tricks include pop quizzes, giving extra-credit assignments, and asking students to solve problems on the board. The list of suggested resources at the back of the book has suggestions about where to find more classroom techniques.

Case Studies

I'd like to single out one particular "trick" because it doesn't get as much attention in the science classroom as it deserves: the case study. This is a time-tested approach to using active learning, with a long history in professional training in fields such as law, medicine, and business. A case study is when you present to students a real-life scenario and then students discuss the resolution of this real-life problem. Case studies are great for group work and focus around solving problems. Some science faculty have embraced case studies as a central part of how they teach, not only for in-class lessons but also for written assignments, as a central feature of a course. You can of course teach using a case study as a one-off, to see if it works for you.

A case study is simply an example tied to the course material. We all teach with examples. We introduce examples for many good reasons—to increase engagement, to illustrate a point more clearly, and to provide real-world relevance for the subject matter. I think it's common for instructors to come up with examples and scenarios off the cuff while teaching. If we see that a concept isn't sinking in as readily as we'd like, we reach into our experience to bring in something more tangible and to give students a more practical handle to grasp the ideas. Teaching with case studies is just formalizing the examples we use, by extending them in further detail and with more planning.

Case studies can be a powerful learning tool because they are built around stories. Research in science communication has shown that a story-based narrative is more engaging with audiences. If you feel that you're having difficulty engaging your students with some content that feels particularly dry, including a case study might be the prescription. If you're teaching an abstract concept that students might have a hard time applying to real-world applications, this is a great time to bring in a case study.

Getting Buy-In from Students

There's a line that I've heard a lot from scientists who are giving teaching advice. It goes something like "It's not our job to make sure students are happy, it's our job to make sure that they learn." Technically, it's true, it's our job is to make sure that students learn. However, learning happens when students are engaged, and if a student is unhappy with their learning environment, it's really hard for them to engage. It's not our job, per se, to get students excited about learning and thrilled about the content. However, doing this will help them learn better, so it's something we should probably pay attention to.

In all learning environments, students are expert backseat drivers. At the end of the semester, students are asked to fill out an evaluation where they are explicitly asked to rate our teaching effectiveness. However, long before that, most folks pass judgment on whether professors are doing a good job and think about what might work better. This is natural.

Getting buy-in from students requires us to convince them that we aren't wasting their time. This is a difficult task. By the time our students have made it to college, they've probably decided that most of their education consists of a large number of hoops that they must jump through to earn a grade. Oftentimes, the hoops that students jump through aren't that closely connected to the actual process of learning. Perhaps the hoops they jump through are an important part of their education, but it might not look like it at the time. When you're in the trenches doing college coursework, it's easy to lose perspective on how all the small lessons add up to a broader understanding. In other words, our students arrive in our classrooms highly jaded about whether what we will be asking of them will actually help them learn. They might be there just for the grade, because what they expect to do for the grade may or may not be really educational.

To get buy-in from students, we usually try hard to convince our students that what we are teaching is important. I think this matters. In addition, I think it's just as important to discuss why our approach to teaching the course is important. Students don't want to know just that the course material matters but also that the particular hoops we are asking them to jump through will help them learn and provide valuable skills.

We learn better when we are conscious about the fact that we are learning, and when we are aware of the methods that we are using to learn. In other words, we need to study metacognitively. Cognition is what happens in your brain when you sort through things and learn. Metacognition is thinking about cognition. If you wander through a maze without paying attention to your route, you may eventually get out, but it will be inefficient and probably unpleasant. However, if you are aware of the fact that you are in a maze and you focus on the methods that you are using to get out of the maze, then you will not only get out of the maze more quickly, but the process of solving the puzzle might be more fun as well.

I think it's critical to discuss metacognition with your students, though you don't need to use that terminology. Learning is far more effective when students are consciously thinking about the process they are using to learn. On the first or second day of class, I am now in the habit of giving a speech that resembles this:

> This semester, my goal is to teach you absolutely nothing.
>
> If I do my job as well as possible, then I will not teach one single fact or concept. Instead, I will set up the circumstances for you to discover information on your own. You only really learn something if you discover it on your own. So, our classes will be set up so that you sort through and find information provided to you, to answer questions and to go through experiences that enable you to make your own inferences and figure out concepts on your own. And you'll be reading a lot outside class.
>
> This course has a destination. We will discuss the routes and there are lots of concepts that interconnect. However, if I hand you a complete set of directions, then you will be deprived of the opportunity of truly learning the way. Don't blindly study the concepts in this course! When you are working on a question or a problem, be sure to recognize specifically that your approach to studying is tied to the larger questions at hand. Please be sure to ask me frequently about the different ways to study the content we are covering in class. Because it's my job here not just to share information with you, but to support you as you work to learn.

Using Office Hours Effectively

I'm familiar with two common complaints about office hours. The first one is "My office hours are empty. Why don't students come into office hours?" The second one is "Right before exams, my office hours are jam-packed!" You can develop some policies to encourage office hour visits throughout the semester. A lot of students, even those who have been in college for a while, have never learned what office hours are. Casey terHorst (associate professor, California State University Northridge) has heard first-generation college students remark that they thought office hours were times set aside for professors to work privately without any interruption from students! You might be required to post your office hours, but that is a minimum expectation. You should also explain that these hours are when you are available to students without an appointment and how they might benefit from visiting. To encourage students to use office hours, some professors have started calling them "student hours." It's worth explicitly mentioning to your class that you welcome visits during office hours.

Office hours are the times that you are available to talk to students outside class hours on a drop-in basis. Beyond that, not all faculty agree about what office hours are for. A lot of professors are glad to spend office hours consulting with students about career plans and to discuss all kinds of academic matters extending beyond coursework. Other instructors regard office hours as an opportunity to deal with student inquiries about course logistics, and still others like to use office hours as tutoring sessions for students.

Most college instructors have office hours that are underutilized, even though many students can benefit from office hours. One way to rectify that problem is to take the approach of Santiago Salinas (assistant professor, Kalamazoo College), who requires his students to sign up for office hours during the first week of class. During office hour visits, he asks his students what they would like to get out of the class, academically and personally.

If you are getting to know a bit about your students so you can help them succeed in (and beyond) your class, it might help to get to know other things that are going on in their lives, especially factors that take a substantial amount of time. In my department, most of our students are working while going to school—some of them are working full-time. During academic advising appointments, I routinely ask my advisees if they have any personal or professional obligations that they think I should know about. At this point, a student might mention if they're caring for a sick relative, or are busy parenting a small child, or have a chronic health concern that might flare up during the semester. I think it's critical to respect personal boundaries when getting to know our students during office hours, and we can ask them in a way that shows we are focused on their academic success. If students are experiencing circumstances that interfere with their academic success, they deserve the opportunity to talk to us about it, so we can do what we can to support their learning.

To a lot of scientists, a packed office before a midterm exam is a positive sign that tests are prompting students to study. In my view, if students are packed into my office before the exam, that makes me worry that I could have handed this situation differently, for two reasons. First, even though I try to schedule office hours to make them available to as many students as possible, there will be students whose schedules do not match up. I think it's unfair to offer last-minute exam prep to some students when others aren't available. Second, this kind of flurry at office hours is often a sign of anxiety, where students are worried about how they will do on the exam, or a sign that students are concerned about anticipating what might be on the exam. Chapter 6 discusses exams, including how we can run exams that don't inspire desperate final-hour office hour sessions.

Using Slides, and How

There are coffee people and there are tea people; there are cat people and dog people; and there are those who lecture with slides and those who don't. No matter how you teach your lesson, you can organize what you are doing using two main approaches: Either you write on the board, or you use a slide show. As is true with much of teaching, you can be a great teacher while never using slides or by always using slides. It just depends on how you go about it.

There are a lot of advantages to using slides to organize your lectures. You can integrate more detailed information than you could use by just writing on the board, and also integrate images and video. Students often want copies of the slide shows that are used in their classes, and you can make them available to students to help them study. If you're using clickers (see later in the chapter), they are designed to be integrated into slide shows. This might be the most effective way to deal with some kinds of complex content, and slide shows with engaging content can increase student interest.

On the other hand, folks who opt against slides have a broad range of reasons, which may or may not resonate with you. By writing on the board, the instructor can help students take notes at a reasonable pace. Slides may serve as more of a distraction. If you give students copies of the slides, some of them may not take many or any notes in class, but if you don't give access to the file, then they may focus on copying the content of the slides instead of cognitively engaging with the content. Some people have argued that using slides can put students in a kind of cognitive tunnel vision, as students have been trained by other instructors to use slides as a template for memorizing information.

If you use slides, then you should prepare for the reality that some students will place an undue importance on them for studying. The late Eldridge Adams (professor, University of Connecticut) used to post written notes from his lectures, but not the actual PowerPoint slides, because he said that "excessive reliance on easily-distributed slides has diminished the quality of instruction. There is a tendency to see slides as the core material that must be learned, outranking even the textbook. It seems to me that some instructors see the course this way too, but putting slides together should not be the primary goal." Some students will focus exclusively on the content of slides as a form of anticipating what you might put on the test, if only because that's what happens in a lot of their other courses. You might be able to make this work to your advantage, by making sure that every lesson has a slide that defines what you expect students to learn from each lesson and that every slide communicates a point that matters.

Regardless of what's in your slides, some of your students will want you to make copies available, even before you teach the lesson, so they can prepare for class. This is, of course, at your discretion. If you don't distribute your slides, it's likely that some of your students will want to take pictures of every slide to help themselves study. Your students are probably accustomed to having instructors lecture for a whole hour using a slide show, and then taking exams that are based on the contents of those slides. When you are using slides for teaching, it might be hard to break students out of that mold. In a traditional lecture course, the most effective use of slides I've seen has been using them as visual accents to the content at hand, mostly with images to illustrate the concepts, with few or no words. This way, they form a supplement to the content, rather than the primary focus of the students. Slides are also useful for providing specific directions for active learning components.

If you're looking ahead to teaching the same course multiple times, having a finely tuned set of slides is both a blessing and a curse. The blessing is that your lecture is already in the can, so you don't have to prep it from scratch. It's possible, in theory, to walk into a classroom with the slide show from the last time you taught that lesson, without having done any preparation in advance. Sometimes it can be pulled off, maybe with just a few minutes of review. It's not a good long-term strategy.

The risk of having a finely honed lecture in the can is that if you keep it long enough, it might develop botulism. When I was in college in the early 1990s, I had a professor who taught introductory zoology using actual slides, in a carousel that mounted on a slide projector. His office had a shelf full of carousels, bearing the slides for lectures for the courses he taught, clearly unchanged for at least a decade. Every day, he'd show up at class at 9 a.m. and pop a carousel on the projector at the back of the room.

His lecture was a well-rehearsed performance, knowing exactly what he would say for each slide. One day, the bulb on the projector was broken, and there was no spare to be found. The folks from AV weren't to be found. So, the professor canceled class. Your slide show is a tool. This tool can enhance your teaching, but it also can become a crutch. If you find yourself in a rut where you're teaching the same class the same way every semester, be careful that you haven't already fossilized into deadwood.

Using the LMS

It's likely that many of your students will be well accustomed to working with the campus Learning Management System (LMS). Your campus LMS might be Blackboard, Canvas, Moodle, or have a cute campus moniker like ToroCourse or TigerLine. Should you use it, and if so, how? You probably have the option of entirely avoiding the LMS, unless you're teaching an online course, which is typically centered around the LMS. There are many good reasons to use the LMS. However, there is one huge potential drawback.

Let's talk about the bad stuff first. It's possible that your campus LMS might be simply a pain in the butt and, as a result, can become a massive time suck. Some of the systems are more intuitive and user-friendly than others. Some campuses provide better technical support than others. If you don't plan to be using the LMS for an extended period in the future, investing time into figuring out how to use it efficiently might not be worth the investment. Secondarily, if the LMS is not well embedded in your campus culture, then it might be difficult to get students to log in as frequently as you might need them to. Using the LMS is a cost/benefit analysis, and if the numerator outweighs the denominator, then don't punish yourself or your students.

There are so many benefits to using the LMS that the activation energy to ramp up can pay off with high dividends, especially if you are teaching a large course and running a technology-rich classroom. There are many features of the LMS, and few instructors of courses avail themselves of all of them.

At the most basic level, you can use the LMS to distribute documents. If you are distributing your lecture slide shows to students, then this is the most straightforward method of delivery. Students who bring their laptops into class should be able to log in directly to download the slide show and make notes during class. The LMS is also an effective way to inform students of their grades. You can use the LMS as your gradebook, so that students can readily find their grades on all assignments so they can calculate where they stand in the course. Integration with spreadsheets is smoother in some LMS than others, but if you want to work with grades offline, you can upload and download them. It is not a bad idea to do a periodic backup of not just your grades but also the full contents of your LMS, as the contents of the LMS are subject to the administration of the educational technology office on your campus.

You can handle all of your exams and written assignments on the LMS as well. You can run paper exams in class and upload the results to the LMS. You also can run a paperless classroom, like Neil Gilbert, who distributes all materials online and receives written assignments from students on the LMS. He also developed self-tests on the LMS to help students study. Going fully digital does has one drawback, as he points out: "The worst part is that it ties me to a computer for even more hours. Grading on the LMS is qualitatively different than grading an actual paper stack of assignments. I don't think either is a hoot."

In theory, you can do it all on the LMS. After all, many online courses take place exclusively with the LMS. If students are taking a multiple-choice quiz by hand (or Scantron), it might not be worth your while to push your exams into cyberspace, but it could be a far more efficient use of your time if your class has 300 students. It can take time to build quizzes online, but once they're set up, the system scales up. Running large exams such as midterms and finals can offer some difficulties, but on the other hand, it's more straightforward for lower-stakes assessments, such as weekly quizzes. If you're teaching a class multiple times, then the LMS can accommodate the construction of a large bank of questions, and you can set the conditions under which students are allowed to retake these assessments.

While the online forum on the LMS looks like it'd be a great opportunity for discussion, I've heard faculty frequently express that the LMS forums rarely result in useful dialogue, and students tend to only post if they are required to. If you're not teaching an online course and are looking to promote discussion, pushing this activity to the LMS might not achieve the engagement you're seeking.

One positive effect of using the LMS is that it can move many of the solitary activities to time outside the classroom, leaving more opportunity for interactive learning in class. Without the LMS, it all happens during class sessions. That means during class sessions, it's up to you to tell students about assignments, have students turn in assignments in class, return feedback on writing, run quizzes and tests, distribute handouts, facilitate student discussion, and give students grades. On the other hand, with the LMS, you can do all of these things with students online, while class is not in session. Depending on your style and the class that you're teaching, this might sound like a cure-all, or like a massive headache.

I see two main factors that can affect whether using the LMS will make your teaching more efficient. The first is the size of the class. If your class is in a huge lecture hall, just the mechanics of distributing, receiving, and managing paperwork can be more pleasant on the LMS. The second factor that can help you decide whether to invest in the LMS is how often you anticipate teaching a particular course. If you do things like create online rubrics, build a bank of quiz questions, and have a rich set of supplementary materials online, then this will lighten the load for future semesters.

Classroom Response Systems

Your institution probably has bought into one of the major "classroom response systems," commonly known as clickers. Clickers are common, and alternative polling approaches involving mobile phones are on the rise. If you're not familiar with clickers, here's the skinny: They're inexpensive devices that record student responses to questions that you ask in class. If you are teaching a traditional PowerPoint lecture, then you can integrate clicker questions into the slide show to check to see if students understand what you've been discussing. The value of the clicker over asking for students to raise their hands is that we are making engagement broader and more accessible.

Used with the simplest functionality, clickers allow you to spontaneously poll the class. This turns out to be more effective and engaging than asking students to merely raise their hands. When integrated into the LMS, clickers can do a lot more. You can integrate questions into slide shows associated with credit for participation and/or correct answers, and you can even use them to give entire exams that are automatically graded in the LMS.

From the perspective of Efficient Teaching, clickers can definitely save you a lot of time. This is true for administering short quizzes and taking attendance. If you want to run a class that requires some level of student engagement, then clickers are a way of engaging students, even in large auditoriums. Clickers can give you a mechanism to keep students accountable for reading, attending, and paying attention, without having to deal with large amounts of paperwork. Once you're in a groove, getting clicker questions ready for class should be far less busy work than processing student quizzes manually using paper.

There are some common gripes about clickers. They're only really useful for multiple-choice responses. (There are some systems that allow text entry, but this mostly triggers headaches.) Some are concerned that clickers may be overrelied upon as an active learning technique, when more impactful techniques are available. If students skip class, they can give their clicker to a friend, so students who plan ahead can still get counted as present. Clickers require you to set up the receiver in the classroom whenever you use them. The software associated with them, which varies among clicker companies, might not be your favorite. Some are concerned about contributing to the proliferation of particular for-profit educational technology companies. Also, clickers are a financial expense for students, though they're cheaper than the typical textbook, and it's possible that most of your students already have clickers from other courses. (Unfortunately, the clicker companies are getting greedy, and some are now charging ongoing fees to keep the devices running after they purchase the clicker. It's possible that your campus has a site license, so that individuals don't have to pay extra.) Ultimately, there are a lot of routes to student engagement, and maybe clickers could be part of your recipe.

Asking Questions

While clickers are a relatively new piece of technology, the real educational innovation underlying clickers is the opportunity for every student to become engaged simultaneously. Perhaps the oldest active learning technique is to simply ask students a question and have them raise their hands. Then the instructor picks a few students to share. When we ask our students a question and seek a volunteer to provide an answer, many of the students don't raise their hands. Some students might, as a general practice, never raise their hands.

When we ask our students to think about answers to questions, we should be asking everybody to think hard. I want my students to emerge from the classroom thinking that they've had an exhausting mental workout. Zumba instructors don't single out individuals to ask them to participate. Likewise, in my classroom, everybody has to dance. When we ask for a show of volunteers, this allows the remainder of the class to shift their brains into cruise control. Over time, the students who don't raise their hands will know that they aren't required to engage. If our job is to support learning for all students, that means finding ways to engage all of the class at the same time.

I know some people who use sets of index cards with student names on them. One approach is to draw names at random from the deck, by shuffling it after each use. Another approach is to set aside the cards you use until you run out, though this means that once a student is called on, they know they won't be required to engage until the stack gets refreshed. Some instructors call on students arbitrarily, or perhaps even students who don't appear to be engaged. This can run the risk of students perceiving (perhaps correctly) that they're being singled out unfairly.

When we are working to broaden engagement to the entire class, we need to be conscious of the well-being of our students. If a student genuinely requires an accommodation because of anxiety, we should work to support their learning in other ways.

Dealing with Digital Devices

I emphasized the necessity of a clear-cut policy about the use of digital devices in chapter 2. Without a clearly stated and enforced policy for exams, if you do intercept a student looking up information or texting about an exam, then it will be far more difficult to levy appropriate consequences (as discussed in chapter 7). To develop the best policy for your course, you need to figure out how laptop and phone use fits in or doesn't fit in while teaching, because there may be times when the issue forces itself upon you. How you make this decision might vary with class size, class activities, content, campus culture, and your own style. There are three general approaches to phone and laptop use in class.

The first approach is an outright ban (with an exception for students who have a relevant disability accommodation, of course). The main argument for the ban is that laptops and phones can be distracting and keep students from focusing on the lesson. After all, if students do have their phones and laptops in use, it seems inevitable that they'll be checking on various social networks and exchanging messages. Another rationale is that students using their devices is distracting to the instructor, who may worry that students aren't engaged in the course content. When an instructor implements a ban, they report that their students are initially unhappy but often come around to appreciate the distraction-free environment as the semester rolls along. To help implement this kind of ban, some professors have used a box for students to deposit their phones at the start of class. This is not an approach that I would use. One huge downside of a device ban is the marginalization of disabled students, by silently making everybody in the class aware of these students' disability status. A related consideration is that not all students who would educationally benefit from such an accommodation will be tested or registered with the campus disability office. Some students who report a preference for the laptop might actually have an undiagnosed educational need for it. It's not our job to evaluate students for disabilities, but it would be nice if we could accommodate all disabled students—including those who are not yet diagnosed. If you are learning toward a laptop and/or phone ban, please consider all aspects of the issue, including the needs of disabled students.

Eldridge Adams used a ban for laptops, tablets, and cell phones in his largest course. He said it improved attentiveness and participation. He added, "The initial motivation was that students who came to class to learn were being distracted or annoyed by students using devices for other purposes. Some students complained to me about others sitting around them playing video games or shopping on the internet and getting rude responses if they ask them to stop. I think it's my responsibility to prevent this, rather than the responsibility of the affected students. Even good students seem to find it hard to forego checking social media during class. Increasing the proportion of active learning methods helps, but not enough. I believe students should come to class expecting to work actively towards the learning goals and that the course should be constructed to support that expectation. After I banned devices (except for clickers), the class focus and mood improved noticeably. The TAs and other observers agree that it's an improvement."

The second approach is the polar opposite of a ban: a wholly laissez-faire approach, allowing devices without limitations (aside from quizzes and exams, perhaps). The arguments for this approach are just as compelling as arguments for a ban. Engaged students may use devices to augment the lesson, by looking up supplemental information. It's been argued that the use of digital devices is merely a symptom of disengagement. In the era before smartphones, it's not as if students were held in rapt attention! They just passed notes instead of texting and doodled instead of web surfing. When a class is built around active learning, then students are often busy enough that they don't have time to be distracted by YouTube or messaging with friends. If students start out the semester with a policy that prevents them from checking their phones, they might feel that they are not entrusted to be in charge of their own learning.

The third approach is a middle ground between a ban and unlimited access. If you are concerned that an outright ban might not be constructive for all of your students, then one pragmatic approach, taken by Sarah Hörst (assistant professor, Johns Hopkins University), is to "divide the room into an electronics-free zone and an electronics ok zone." This approach seems to have all of the advantages of a ban for some students but avoids the negative effects of the ban for others.

As you develop your own policy for dealing with devices, I will emphasize one important point, rooted in the Respect Principle: University students are adults, and actions that constrain their liberties can get in the way of building a rapport. You might feel that your students are still sub-adults, still dealing with how to manage their own independence. Regardless, the more you try to control behavior, the more your command of their attention will slip through your fingers. If we choose to use our classes as a digital abstinence clinic, this has the potential to harm the learning environment more than help it.

My own approach is to encourage students to avoid devices unless necessary. As I say in my syllabus: "If you have a need or preference to use a laptop, that's fine. Please avoid doing things that aren't related to the class. If you do use a device, you might be asked to sit in a particular location in the room that I think is most suitable for the learning environment for other students in the class. Audio or video recording in class is prohibited unless prior authorization is granted. No devices are allowed during assessments unless specified otherwise." If I find a student busy on the internet doing things unrelated to class, I try to not take it personally. I understand this is about them and not about me.

Formative Assessment

Teaching is more effective with frequent assessment. This is no surprise, and this is one of the reasons why Efficient Teaching is not effortless teaching. We can help students learn by making sure they get feedback about whether they're getting it right, early and often. Knowing that feedback helps students do better, we need to seriously consider how we can make students sure they are on track before they take major exams and turn in major assignments. When we give students a regular heads-up about where they're learning and not learning, this "formative assessment" is effective pedagogy. It involves more work on the part of the instructor on the front end, but the work pays off because students are expected to pick up the material more quickly and with fewer interventions required.

Formative assessments can take many forms. It makes sense that your choices will depend on the content of the course, the size of the class, how you've structured the curriculum and teaching methods, and customs in a given institution. I'll walk through a sampling of approaches to frequent assessment.

Some of the most-established teaching elements are employed as formative assessment: quizzes and homework. When students get their scores, this will let them know how their work is (or isn't) paying off. If they find out which questions they got right and wrong, this can help them plan more effectively.

If the formative assessment involves grading by the instructor, then it gets more difficult to do as classes grow in size. The best formative assessment provides specific and prompt feedback. If you're buried in a ton of grading, then it's hard to do this well and to do it promptly. How can you manage this quantity versus quality trade-off in the long term?

It is possible to make sure that formative assessment happens without having a massive stack of grading that constantly taunts you from your desk. If you are teaching a lecture to several hundred students, then this is often the kind of thing that gets assigned to teaching assistants! Of course, not many of us have an audience that is huge enough to warrant that kind of personnel support.

While instructor feedback makes formative assessment more valuable, students don't necessarily need your marks on their paper assignments to get useful feedback from their work. One approach is to give regular ungraded quizzes in class. It doesn't take that much time to administer a five-minute quiz. Instead of asking students to turn in their quizzes for grading, you can simply ask them to grade their own quizzes (or pass them to a neighbor, at their own discretion). By giving them the answers right after they finish the quiz, that's about as immediate as you can get with feedback! This also gives you feedback on your own teaching, to see what is connecting and where you need to spend more time or change to a different tack. You can do a similar thing with homework—you can assign homework to students and have them bring it into class, and then you can go over the homework together in class. There are many different ways that you could think about how to assign grades to this, or decide to not assign grades. Some instructors who are concerned that students won't take ungraded assessments seriously will randomly grade a fraction of the assignments, without announcing in advance which ones will be graded. I caution against this approach because this is likely to generate an environment with consistent low-level fear, but you can temper this by making the point values of these assignments very low or by dropping several of the lowest scores.

One of the biggest utilities of instructional technology has been the ability to scale up formative assessment without increasing instructor workload. When you run a quiz on the LMS, clickers, or with mobile polling, it takes essentially the same amount of work on your end, regardless of whether the class has 40 or 400 students. If you've decided to put in the front-end investment to get this ramped up for your course, the more often you run these assessments, the more you can streamline your workflow. If you've integrated clickers with the campus-wide system for the LMS, then assigning points for answering (or for correct answers) for in-class assessments is straightforward. If student interaction is not an important element of these questions, you can run these quizzes on the LMS.

There are serious limitations about doing formative assessments using instructional technology. The trade-off for the instant feedback is the limited range of questions that you can ask. The system operates best if you use multiple-choice questions. In principle, you can use short answers, but if the student's answer is correct but doesn't perfectly match the short answer you specify, they could be marked as incorrect, and this becomes a big problem at scale if you are assigning points for correct answers. (For example, if the correct answer is "12" and a student writes "twelve" or inadvertently puts a space before the 1 or after the 2, it's possible they could be marked as wrong.)

Textbook publishers have developed some sophisticated online portals for digital learning that can be tailored to fit the curriculum that you're teaching. Unless you have a team of tech people who are also content experts backing up your course, it's likely that these textbook-based online learning-and-testing systems are going to be slicker, less bumpy, and easier to navigate than anything you could cook up in your own LMS.

There are two major disadvantages of adopting the textbook's formative assessment platform. The first is that you have abdicated the decisions over the relative importance of content over to the publisher. If you think the textbook is so spot on with your learning objectives that you can just let the textbook be the source of exam questions, then this won't be a problem for you. However, many instructors will have learning objectives of more specific design and might be concerned that the test material simply gets drawn from the publisher's test bank.

The second concern is that these systems are very expensive, often costing several thousands of dollars per section. These costs are entirely passed on to the individual students, who must purchase a license to access these proprietary materials along with the textbook. While some publishers work to find a price point that is not too painful to students, you may want to consider how much money students are handing over. For example, consider a nonmajors biology course with 100 students, and the cost of the online learning platform is $50. The company is receiving $5000 from students to operate this platform—this is often the cost of employing an adjunct instructor to teach a whole lecture course from scratch! After this section is taught for four semesters, that's $20,000 of money from students that, in theory, could have been applied to the development of educational materials that meet their particular needs in your particular institution.

There are many ways of providing formative assessment for students that aren't addressed here. In the notes at the end of the book, I suggest resources that will provide you more ideas than you could possibly implement. Feel free to rely on the wisdom of people around you and experiment with new approaches as well.

Keep in mind that "active learning" in the classroom typically involves formative assessment. If you're building more active learning elements into your course, then the formative assessment pieces can more easily come along for the ride.

Learning Styles Are Not Real

This might come as an unwelcome piece of information, but I have to break it to you that "learning styles" are not real. Some people say they're "visual learners," or "auditory learners," and so on. There doesn't seem to be much empirical evidence supporting this theory. Educational research has shown that there are no individualized differences in how effectively people learn based on the format that information is shared. I realize that "learning styles" is a commonly accepted popular notion, but it's one of those urban legends that just doesn't stand up to scrutiny.

If someone tells you that they have a certain kind of learning style, this isn't something about how their brain works, or even about how they best learn. It just tells you what their preferences are. It's up to you to decide how many student preferences you wish to accommodate, of course.

Just because learning styles are not real, this doesn't mean the mode of information delivery is unimportant. Some topics might be taught better in some formats than others. I would imagine that using videos to teach introductory mechanics in physics would be rather impactful; images of organisms are important for biodiversity; and long narratives might be useful for other subjects. Keep in mind that some of your students will have disabilities that will require accommodations involving mode of information transmission. These required accommodations aren't designed to support a person's style but, instead, to provide educational access to students with a diagnosed disability.

The fallacy of learning styles doesn't give you license to teach in a single modality, because the complementary use of different mediums does help people learn better. The use of video clips, images, stories, conversations, and drawings at the whiteboard all can combine together to increase learning. But rest assured, if you don't use every mode equally, you aren't conferring a cognitive disadvantage to any of your students based on their professed learning style.

Lab and Field Safety

Safety always comes first. Working in the lab and working in the field presents many kinds of safety hazards. It's our responsibility to be aware of risks, reduce the probability of an incident, and prepare ourselves and our students in the event of an incident. Our institutions should require safety training for any work in the lab or the field, including the completion of paperwork to document that individuals have been trained. These trainings often are pro forma and conducted without much care. Please train your students comprehensively.

We take safety seriously by enforcing safety guidelines. If we see a student violating a safety rule, then we need to take immediate corrective action. If the rules say you need closed-toed shoes in the lab, that means you can't slide on the rules and allow a student to attend lab while wearing open-toed shoes. If the rules say no food in lab, then you need to immediately put a halt to eating in the lab. If the rules say that students need to have water with them when going into the field, you need to make sure everybody has water before you depart. If students pack into a vehicle tightly enough that not everybody has a seat belt, then you need to come up with an alternative arrangement so that everybody has a seat belt. You get my point. By taking safety seriously and as a prerequisite to learning, you make sure that people follow the rules that keep them safe.

TAing a Lab Section

If you're like most scientists, then your first experience with college-level teaching was as a graduate teaching assistant, which I'll just call being a TA. "Teaching assistant" sounds like you'll just be there to help out, but if you're teaching your own lab section, that typically means that you're the "instructor of record" for this lab section. This means that you're the one who assigns the grade that ends up on the report card, and you're ultimately responsible for what happens in the lab. As a TA, the classroom is yours. If you think this is an astoundingly high level of authority to give to a person who may have no training and no experience in teaching, then I agree with you!

If you're teaching a large introductory course, then there will be many lab sections that are organized by a coordinator. The labs are typically set up in cookie-cutter fashion, so that anybody can step in and teach, after a brief orientation from the coordinator. I think these teaching assignments might be one of the hardest ones in higher education, even for instructors who have a lot of experience. This is because you're responsible for something that is not entirely your own creation, and you may have little control over things that are important to student success. Of course, teaching these lab sections may be easier in the sense that you don't have to prepare the lessons from scratch. However, without full control over how students are assessed, you're not fully empowered with the authority to help students succeed. If you're going to help your students perform well, you'll have to accept what you have control over in the lab and what you don't control. Then you can develop your teaching to provide students with what you anticipate they will need to perform well in the parts that you do not control.

It's likely that the lab you're teaching will have some kind of canned curriculum. You'll need to understand the content well enough to answer basic questions and be prepared to seek out answers for more sophisticated questions. In a lot of disciplines, grad students and professors may not be prepared to navigate the waters of introductory curriculum. If you're teaching a physics lab, can you solve a problem using Ohm's law like you could as an undergrad? In biology, for example, introductory students typically learn the classification of organisms far better than most grad students will know off the top of their head. You may have learned about the Apicomplexan protists that cause malaria briefly as an undergrad, but you probably should bone up on this before teaching it!

These lab sections become a bigger challenge if you don't have control over the grading of student assignments or exams. It's possible you can decide what students will be turning in for a grade, whether it's a completed lab handout, or a lab book, or a lab report, or a quiz of some sort. Typically, you'll be in charge of grading these. There is a greater risk, however, that the students may be taking exams that you are not in charge of writing. You might not even be able to see a copy of the exam before you're teaching! This puts you in a tight spot, because you will be responsible for preparing students for an unknown target. It's hard to avoid developing inappropriate expectations about what students are supposed to be tested on when you didn't design the activity yourself. Keep in mind that your knowledge might be contradictory with information that students are getting from the lab itself. Of course, be sure to teach students the correct information. In general, when you're not the only one developing an exam for your students, it's a good idea to get as much information possible about how the exam will be developed, before you teach the content.

Inevitably, you will be experiencing some bumps because you're teaching someone else's course. Every lab activity will have problems, no matter how well it is designed. The results might not come out as you anticipated, or you might not have enough materials, or some parts of the lab manual might not make much sense. And yet, as the instructor of the lab, you are wholly responsible for this. It might be tempting to disavow some level of responsibility for the lab, to say something like "I'm sorry that not everything is working out so well. I didn't write this lab, but I'm glad to be here teaching it." This is honest, but it's not helpful. Throwing the curriculum or the lab coordinator under the bus won't create a positive outcome for yourself or for the student. As far as students are concerned, you are the person in charge, and when it comes to their grade, the buck stops with you. It might be uncomfortable to take ownership of a curriculum or an exam that's not yours. This is better than making your student uncomfortable by making remarks that make them less confident in the instruction that they are receiving.

Individualized Instruction in Lab

My first semester in grad school, I taught an introductory lab that ran 64 sections each week. I was one of two dozen TAs for these sections. Each Friday we'd take down the lab for the week and set up the lab for the following week. We all used the same lab manual; students were required to turn in the same lab report forms each week, and they even used the same final practicum. Despite the remarkable similarity among lab sections, it didn't take long for me to realize that everybody taught their own section rather differently, especially the grad students who had taught this course in previous semesters.

If you're running the lab section, then you have substantial control over what happens in the lab. You're the one grading the lab reports, presumably with the grading criteria that you've specified to the students. Students don't immediately dive into lab activities, so you can give them a lesson of your choice at the beginning, to contextualize the lab and teach what you think is most important. The more you make the lab your own, the more you can adapt your approach to meet the needs of your students.

Lab Partnering

Most labs are structured for students to work in pairs or larger groups. How groups pair up can be important, because discussions among partners are how problems get solved and how people learn. Groups can also be a source of conflict, and a lot of difficulties of students come from issues among lab partners. To prevent these issues, it's good to have a solid set of policies for dealing with lab partners from the outset. You can choose to designate partners or let your students self-select. Rather than reinventing the wheel, it's a good idea to know what other TAs have done in the past and what is thought to work best. It would be helpful if you were to generate a list of what you expect students to be doing with and for lab partners and what you expect them to be doing independently. Are lab partners supposed to record one set of data and share, or is everybody supposed to record their own data? What parts of the lab assignments do you expect to be a group effort and what do you expect them to do independently? Is any part of a student's grade going to be derived from the assessment of the contribution by one's lab partner? Do you expect students to keep the same set of partners every week, do you expect them to mix it up, or are you not particular about this? There are a range of valid ways to answer these questions, but making these policies clear upfront can make it easier down the line. Group projects are discussed in the next chapter, and general principles about group work often apply to how we run lab sections.

Lab Attendance

Lab attendance issues are radically different than attendance at lecture sessions, so whatever policies you use for lectures shouldn't necessarily be applied to labs. For many courses, a student could skip a bunch of lectures, but if they do all the assignments and perform well on exams, then they could probably pass just fine. However, the point of being in lab is to actually perform the lab and report on it. Because repeating a lab is not feasible (or, if it is, would require plenty of extra work on your part), it's hard to deal with attendance problems after they occur. Having clear attendance policies at the outset is particularly critical for labs.

Another attendance problem unique to labs is that if a student shows up late, they might not be able to join the activity if the class has already started. I once taught a nonmajors lab that started at 7:30 a.m., and as the semester went by, more and more students arrived later and later, to the point of absurdity. I could have prevented this by making it very clear from the start that students need to arrive at lab on time, but I had not anticipated this problem at the outset. I could have created a policy such that if someone showed up too late (say, 15 minutes past the start time), then they would not be able to receive credit for the lab. I'm not a fan of no-tolerance policies, but having students arrive at lab well after it started helped nobody. If you are planning to drop the score for one or two of the lowest labs, then showing up late shouldn't hurt the student unless it develops into a habit.

Another issue with lab attendance is that sometimes a student might get weary and will slink out without bringing attention to themselves. I've taught some labs where I am focused on working with one group of students, and then I look around the room, and—poof!—a few students have left prematurely, leaving behind their lab assignments to be graded. Their labs were incomplete, and they would end up earning a poor grade. Students that are not comfortable or confident in lab might be inclined to leave prematurely, which can feed a positive feedback loop resulting in lower performance. You can support continued participation in lab by checking in with all students on a regular basis. If some students look like they're doing just fine on their own, it's still a good idea to see how they're doing, ask if they have any questions, and do a check for understanding for the activities they've been working on. It's also a good idea to require an "exit ticket." If students are turning in what they worked on, having them hand it in to you personally is better than depositing assignments on a lab bench. If students are turning in their lab product at a later time, then you could ask them to answer a short question on a sheet of paper about what they learned from lab, or a question they have that you could answer in the next session.

5

Assignments

Homework

We usually expect students to be able to apply what they've learned in the course to solve problems. Moreover, solving problems is a way of reinforcing what students have learned. I am a fan of assigning problems for homework because this provides an excellent guide for studying effort. If we ask our students to just read the book and study material that is discussed in class, then many students will have difficulty knowing how to study to solidify and expand their knowledge. On the other hand, if you have a set of problems that you would like students to solve for homework, then working on these problems can help them direct their studying effort in the direction of your design.

If your course uses a traditional textbook, then a bunch of questions and problems will be in each chapter. These might be excellent homework assignments, and you can pick out the questions that articulate best to your objectives. In theory, the "practice" questions have answers at the back of the book, and "assignment" questions have answers in a key that is only made available to the instructor. In practice, answers to all of these questions are readily found online, in legitimate or gray-market websites. If you use these questions as homework, be aware that your students have the option of tracking down these answers. The students who cheat on this homework are only cheating themselves, but it's our responsibility as instructors to create a learning environment where the shortcuts are not readily available. Especially if homework assignments are a big part of students' grades, developing your own questions would be a good idea.

Grading homework can get onerous very quickly. If you are merely checking to see that students did the work, then you can breeze through it more quickly than making detailed remarks. If you provide students with copies of solutions to the homework problems, this isn't as excellent as individualized corrections, but they still can use those solutions as a study guide.

Grading Rubrics

When grading, we assign points for correct responses. With assignments that have more than one single correct answer (for example, lab reports, essays, multistep problems, in-class presentations), we are assigning points based on multiple criteria. We have to keep track of the criteria we are using and their relative weight, to make sure we are consistent and fair. When you formally write down these criteria, you're making a rubric. A grading rubric is simply a breakdown of the points that you are adding up to score an assignment. When you're grading, you should know what you are looking for, and having that rubric formalizes those criteria.

If you're handing out an assignment that is going to be evaluated with a rubric, students deserve to see this rubric as part of the assignment. For example, when students are turning in a term paper, you can let them know what their grade will be based on. In my opinion, if you're working from the standpoints of the Respect Principle and Efficient Teaching, then using rubrics for written assignments is an obvious slam dunk, because the alternatives to rubrics lack transparency and don't guide students toward what they need to do to succeed.

Let me describe how grading might go without a rubric: Let's say you're grading a lab report. You have a good idea what differentiates A-quality work, B work, and C work. Clearly, there are a lot of factors involved, including the level of detail, whether the work is factually correct, whether it's well written, whether the student followed instructions, and so on. As an instructor, it's entirely within your rights to look at an assignment and holistically grade it. A perfect lab report comes in: full credit! A lab report comes in that is missing a few important details, but there are no factual inaccuracies, and there is a missing table that you required. Call that a C, 75%. This is how a lot of instructors grade: They tell students what is required, and if the student doesn't do it well, then they don't get full points.

If you use a rubric for grading lab reports, then you give students a breakdown of how they earn points on a lab report. Let's say out of 100 points, you designate 20 points for each of the five categories: adequately detailed description of the methods; figures and tables as required; scientific and mathematical accuracy; accuracy and depth of discussion; clear and logical description of the hypothesis in the introduction. Of course, you presumably have different expectations of your students in your own lab section, so you can structure your rubric to reflect those expectations.

When you give students a grading rubric when you hand out the assignment, this shows respect for their time and effort. Writing anxiety is a serious concern for many students (and many of us, including myself). One of the best ways to relieve anxiety about writing is to make the target as clear as possible. By making it as clear as possible upfront how their grade is calculated, they will be more confident that they can hit the target that you set. Students will look at the rubric as they are writing and will not have to worry as much about what you are looking for.

I understand that some folks don't like using rubrics because they think written assignments should be evaluated holistically or by gestalt. I think it's possible—at least in theory—for a holistic grading system to be used that can provide a fair measure of student performance. However, the drawback to this approach is that it's not student centered. Even if it's possible to grade written assignments without rubrics, students who are working on the assignment will not have any concrete guidance about what to do to make their writing more effective.

Whatever you think students should be doing in their assignment, you can put that as a component of the rubric. Is bad grammar something that deserves points off? Put it on the rubric. Should it be impossible to get an A without a clearly articulated thesis and well-supported arguments? Build that into the rubric. Does citation format matter to you? Put it on the rubric! Don't care about citation format? Then don't put it on the rubric.

Rubrics make sense in terms of teaching efficiency. They save you time before grading, while grading, and after grading. With a detailed and specific rubric, you are less likely to be approached by anxious students wondering about what you're looking for in the assignment. If they do ask you specific questions about the rubric, this gives you the opportunity to engage students about the substance of the assignment, rather than mere grading logistics. Rubrics make the grading process easier, as you can organize the process into discrete pieces, and you do not have to write as many detailed comments, as the rubric will inform students which parts didn't earn full points. After you hand the assignment back to students, students will have a straightforward explanation about how they earned their grade, and any questions that students might have will be focused and more constructive. If they see exactly where on the rubric they lost points, they are far more likely to use their own time to figure out what they need to do to improve their performance, rather than hassle you about it.

Most importantly, rubrics result in better writing practices from your students. It is a rare student who relishes receiving a draft of an assignment with massive annotations and verbose remarks about what can be done better. Those remarks are, of course, very useful, and students should get detailed remarks from us. When fixing the assignment, students will be focused on getting a higher grade than they received on their draft. We promote student success when we let them know the specific categories on which they lost points. This kind of diagnosis, along with any written comments that professors wish to share, is more likely to result in a constructive response and is less likely to terrify students who are unclear how to meet the expectations of a professor who gave a bad grade without providing a specific breakdown about how that bad grade was assigned. If a student wonders, "What can I do to produce excellent writing?" all they'll need to do is look at where they lost points on the rubric. That's a powerful diagnostic tool. If you think the use of a rubric in your course cannot be a great diagnostic tool, then you haven't yet designed an adequate rubric.

Group Projects

Working collaboratively is considered to be a "high-impact educational practice" and also a "twenty-first-century skill." It's unambiguously great for our students to learn while working on collaborative projects. After students graduate, they will be expected to work in groups regardless of their professional endeavors. Students who are not working in groups in college are, arguably, being shortchanged.

It's also an unambiguous reality that group projects often turn into a minefield for students and instructors. I recently saw a witty unattributed quip (a meme, you know) that said, "When I die, I want the people I did group projects with to lower me into my grave so they can let me down one last time." Whether you assign groups to students or allow them to pick their own group members, you'll have people unsatisfied with the composition of their group and resentment about an unfair division of labor. You'll probably be called on more than you would like to deal with problems involving group dynamics. Especially if students aren't used to working in groups, the work will end up getting done at the last minute and the quality of the work might not meet your expectations.

If you are planning on assigning group projects for students to collaborate on outside class sessions, I recommend careful planning and monitoring, and also consulting with other instructors in the department who have experience running successful group projects. For more concrete approaches, I suggest looking into recommended best practices from the "team-based learning" approach. Group projects rightfully have a reputation among students as an overall negative experience. If you're planning to use group work as a part of your course, it's time to do some homework on established best practices.

For starters, here are a few suggestions to help group work go smoothly. Most problems come from nonconstructive group interactions, so how you assign groups is critical. If you let students pick their own groups, some of them probably won't work well together. Randomly formed groups might be better, though assessing students to create complementary groups is considered to be the best approach. How you form these complementary groups may depend on what the group is doing and the backgrounds of the students in the course. You could make sure that each group has a student who you can identify as interested in relevant tasks, such as recording data, using a spreadsheet, making graphs, or editing video. In long-term groups that work together closely (such as in problem-based learning), it would make sense, when administering an assessment of knowledge and ability, to group students so each group has complementary knowledge and abilities.

It is important to keep students accountable for their teamwork. You can do this by having students assess one another's relative contributions and teamwork. Frequent feedback can keep groups working more smoothly, so scaffolding the assignment so work is due in steps is helpful. It is most common for groups to turn in a single final product and for everybody in the group to receive the same grade. Some instructors opt to give some weight to student evaluations. For example, the approach taken by Sarah Lawson (assistant professor, Quinnipiac University) is to assign 75% of the grade for overall group performance and to reserve 25% for peer evaluations.

If you're not up to managing group projects in your class, that's fine. There are a number of ways your students can gain experience learning and working in groups without assigning a group project. You can use group-based approaches inside your regular lecture and lab sessions and can also encourage students to work collaboratively on assignments that they turn in individually.

Term Papers and Major Reports

Science classes tend to lean toward lab reports and short essays rather than a single large writing assignment that can count for a substantial fraction of the grade. Because a lot of the writing that scientists do is in a shorter format, this is a skill we need. We also write major reports, research papers, and reviews, all of which require substantial literature research. If your course involves a major paper that students are expected to turn in near the end of the semester, I want to highlight two recommendations: provide students with the grading rubric at the start of the assignment, and scaffold the assignment with intermediate steps.

If you have high expectations for a major written assignment, then you want to be sure that your students understand these expectations and will put in the effort required to meet them. The most direct way of doing this, in a way that will banish ambiguity, is to provide a detailed grading rubric that specifies your expectations for this assignment. I've already provided a detailed rationale for grading rubrics; I just want to emphasize that for a major project that is supposed to take many weeks of sustained effort, providing students with a detailed target is important.

Scaffolding writing assignments is good pedagogy, because good writing requires multiple steps and revision. This experience can help all students. A scaffolded writing assignment includes smaller assignments that students turn in, as a process of building up to turning in a final complete draft.

Scaffolding writing assignments is friendly to students because it reduces the cost to students who end up working on major assignments at the last minute. If you're interested in teaching the process of writing, then providing feedback to students on the intermediate stages of work is important. People procrastinate, and this is particularly true with written assignments, often because we tend to underestimate the amount of work it takes for us to generate an excellent and well-researched piece of prose. If you don't require any kind of preliminary work to be turned in before the final paper, some fraction of your students will be putting this off until the last minute. In this case, giving them feedback well in advance can help them space out their time. Since you are presumably asking students to delve into some questions of substance, as they are developing their projects, they will benefit from receiving feedback from you. Scaffolding provides a formal opportunity to steer students into a constructive direction and provide them support. If you are counting on all of your students to get quality feedback by approaching you for help, you will not be able to reach out to as many students, and for those who do come see you, your feedback can be a springboard for a more rewarding discussion.

There are many kinds of intermediate assignments that you can require prior to the final draft of a major writing assignment. These provide you with the opportunity to give feedback to students to let them know if they are on course, or to steer them in a more fruitful direction before they get too far off track. The first assignment can be merely a suggested title and a short description. If this is already developed, then an annotated bibliography would be the next step, in which the student shows that they have gotten well into the literature of the topic and have identified a number of sources and demonstrated how these are informative. Other steps can include an outline of the paper and a first draft of the assignment. A paper from laboratory or other research can be divided by sections, introduction, methods, results, and discussion. You also can require students to turn in multiple revisions. Consider formalizing a system for students to exchange papers to assist one another in the revision process, though the quality of edits will increase if you evaluate these edits.

Plagiarism and Online Plagiarism Checkers

Plagiarism is widespread in academia and beyond. While most university instructors have been well acculturated with a set of community norms about what constitutes honest scholarship, some of your students may not be as familiar with these community norms.

Why does plagiarism happen? In some cases, students genuinely do not understand the distinction between proper research and copying and pasting. Alternatively, some students don't understand the distinction between paraphrasing a statement in one's own words versus tinkering with the wording of something someone else has written. Many students arrive to college with little experience with—or accountability for—formal academic writing.

Your institution probably subscribes to a service that can screen student assignments for plagiarism. If you're not familiar with these tools, they cross-check the text of student assignments against the text of files broadly found on the internet, as well as peer-reviewed literature and books and also a database of any other papers run through the system submitted by your students or anybody else who has used the service. (It is possible to ask them to exclude your students' work from the database, though given the privacy history of ventures such as this, one has to wonder how this might make a difference.) The upshot is that you get a report with some percentage value showing various strings of text that match up against other pieces of writing.

Before we go any further, how does the Respect Principle relate to online plagiarism-checking tools? One valid perspective is that plagiarism-detection software interferes with a trusting learning environment. For example, Jesse Stommel (director of teaching and learning technologies, University of Mary Washington) does not use plagiarism checkers. He wrote on Twitter, "Students who understand the what and why of the tool have expressed (to me) that when they are made to use [plagiarism-detection software] they feel they are presumed to be cheaters. What they want is an environment of mutual trust and respect." I'm not recommending that you use plagiarism-detection software by default for all of your students. I think the more we emphasize the adversarial nature of detecting misconduct, the harder we make it to focus on the concepts and practices of the course itself. We only have so much mental energy we can dedicate to our teaching, and expending it in this direction might do greater harm than good. However, I think I should expand on plagiarism-detection tools, particularly for science classes, as this is a common practice.

Plagiarism-screening services can be used in two different capacities: prevention and detection. If your students are turning in written assignments through the LMS, then you likely have the option of creating an assignment that gets automatically screened. You will have the capability to enable students with access to the plagiarism detector so their assignments are screened in draft form, before they turn in the final version. This allows students to check their own writing for plagiarism and to fix the problem before they turn it in to you.

Using a plagiarism detector as a device to educate students about plagiarism has upsides and downsides. If a student plagiarizes without being aware that they have done so, then this is a way for them to discover this without turning in a piece of academically dishonest work. While a lesson in honest academic writing is the prescription to prevent plagiarism (as discussed in chapter 7), this is a different way of empowering our students. Then again, if a student has made the choice to plagiarize, then asking them to filter their work through plagiarism software will only serve to help them circumvent detection in the final product. If your goal is to teach honest academic writing, then this might be part of the equation, but the education has to start before the draft gets written.

If your goal is to catch the cheaters, then the detection software indeed will catch a sizable fraction of the plagiarized assignments. If this is your disposition, then at the very least your students deserve a full lesson in what constitutes plagiarism and how to write without plagiarizing. What is the success rate of this plagiarism-screening software? That depends. There is an ongoing evolutionary arms race between the developers of the detection software and those who submit assignments with a goal to subvert its effectiveness. There will always be plagiarized assignments that sneak through. Faculty who regard themselves as quite capable of detecting plagiarism often fall short in this ability. If you feel that you just have the capacity to recognize plagiarism when you see it, please know that you're probably wrong. I am basing this assertion on the very high incidence as reported in the academic literature, relative to the frequency with which students are being detected by their professors. And, of course, there are plenty of independent contractors who market themselves to students, who will create bespoke written work that will get a green light from the plagiarism-detection software, which has been called contract plagiarism. Ultimately, relying on the capacity to detect plagiarism is not a panacea for dishonest academic writing.

Scaffolding written assignments is solid pedagogical practice that will also reduce the incidence of plagiarism, because students will feel less motivation to plagiarize throughout the process. It is possible for students to avoid performing original work on a scaffolded assignment, though it would be more difficult and less rewarding. Chapter 7 has more detail about preventing plagiarism and dealing with it once detected.

Extra-Credit Assignments

I periodically see an old quip: "How do you get an undergraduate to cross the road? Extra credit!" It's a mystery of human nature to me, but there are many students who apparently go to great lengths to earn a few extra-credit points but who do not put as much time or effort into earning points for regularly assigned work. The idea behind extra credit usually is that it increases student motivation.

Some extra-credit activities simply are typical class business that aren't included in the regular points scheme, or aren't mentioned on the class syllabus. These things include rewriting assignments or retaking tests, or some kind of homework activity. This isn't "extra credit"; it's more a form of spreadsheet trickery, so the total number of possible points exceeds 100%.

However, some "extra credit" is designed to not be equally accessible to all students in the course—if they require participation in an activity that falls outside regular class hours. This might involve attending an event on campus or participating in some kind of volunteer activity. In my opinion, using this type of extra credit creates more problems than it solves, and it falls short on the Respect Principle.

When students sign up for our classes, they are expected to attend class at the scheduled times and to complete the studying and assignments outside class, though not at any particular time because they have other courses, jobs, and private lives. The syllabus says what is in the class and why the class exists. If a professor adds additional stuff to the course, at some point through the semester, then this provides a disadvantage to the student who has more commitments outside regular class hours.

I understand that a lot of unexpected things can happen during a semester. I'm all in favor of an instructor deciding to alter course to respond to changing conditions, as long as it doesn't violate the policies set forth in the syllabus. However, I think it's critical that whenever you are offering additional points to students in the course, all students have an equivalent opportunity to earn those extra points, without requiring more effort from some parties than others. For example, if you offer extra credit for students to attend a special seminar, then students who have other duties in that time slot are put at a disadvantage. If you assign an essay to those students who can't make the seminar, that's a punitive measure.

While not offering extra credit might make you seem less happy-go-lucky to students, it does give them the impression that you deal with them honestly and fairly, and do not do (seemingly) capricious things to change how grades are calculated. If grades in the class are lower than you intended, because it was more difficult, it is more fair to shift your grading scale for all students than to offer extra credit to bail out students with lower scores. You are likely to have some students approach you for an opportunity to earn extra-credit points, particularly near the end of the semester. It would be unfair to offer this opportunity to those students without extending the same opportunity to all students in the course.

Lab Assignments

Assignments for labs can be all over the map. I presume this is because what we expect students to learn from labs can vary substantially. What kind of experiences from your lab course will they need to take into their next courses, or beyond graduation? Do you want them to demonstrate that they learned a set of facts and concepts from labs, or do they need more experience with writing? Have your students learned how to keep a proper lab notebook, and should you be evaluating them? Are students expected to develop and demonstrate proficiency in certain laboratory techniques? The answers will depend on other courses in the curriculum, to make sure that your lab fits into the overall educational goals for students in the program of study.

Lower-division courses often have cookbook laboratories in which students are expected to fill out handouts by completing lab activities. These handouts may involve labeled drawings, problem-solving, responses to more open-ended questions, the creation and analysis of datasets, or providing the results of an experiment. Filling out these assignments helps students learn by providing accountability for the activities that are expected of them in lab. If students need knowledge and an opportunity to gain proficiency in basic laboratory activities, then perhaps this is what you need to implement in your lab. You could include all of these assignments in a single volume that students receive in advance of the lab.

The student lab book could simply be a composition book that they are required to turn in at the end of lab every week, to be graded and returned to the students for the start of the next lab. This is most common when the lab activity is designed for students to gain experience recording how steps were taken, for instance in an organic chemistry lab.

Another type of lab report is a formal write-up conducted after the lab activity is over, turned in at a later time. This often follows the format of a brief scientific paper, with introduction, methods, results, and a discussion. This is expecting a lot more work from students—and if it's not necessarily more work from your perspective, it nonetheless will be considered to be a more substantial assignment by your students. This kind of lab is often done when labs stretch over multiple weeks, so that a smaller number of lab reports are turned in over the course of the semester.

There is great inconsistency in the format of lab reports from course to course, and there is a lot of variability in what lab instructors expect to see in lab reports. For this reason, it's particularly important to communicate your expectations upfront, which can be as simple as providing a grading rubric for all lab reports. You might consider handing to students models of lab reports that meet your criteria for earning full credit, so they know what to shoot for.

6

Exams

Do You Need Exams?

Exams are a long-standing tradition, and they are a big deal. As you can see, here is an entire chapter about exams. Before you dig into this chapter, I'd like you to ask yourself this question: Does your course need exams? You might not have considered the possibility that you can teach without exams. If you weigh the pros and cons, do you think it's possible that your students will learn more if you don't use exams? Let me walk you through some upsides and downsides, and uses and misuses, and you can decide what is in the best interest of your students.

We have big exams because this is what we're expected to do. Whenever your teaching goes against institutional expectations, it might be seen as a negative by peers who do things differently than you. You might not even have the latitude to choose against exams, if you're teaching one section of a lecture or a lab that is designed to match up with other sections that are being taught by other instructors. If you happen to be in an environment where conformity is valued, then going exam-free might not be in your interest in the long run, especially if it ends up being unpopular with students (or perhaps too popular with students).

Exams are principally used for "summative assessment"—not a way-point, but an instrument to measure how much a student has learned before moving on to a new stage. This kind of information is appropriate for assigning a grade at the end of the semester. Giving exams is a pragmatic way to create the data you need to assign grades. Summative assessments can be useful not just for you, but also for students. In the middle of a semester, it's possible that a big summative assessment (also known as a midterm) can be a wake-up call for students who aren't performing well on smaller assessments. In this sense, the grades on exams are an extrinsic motivator to encourage student investment in schoolwork.

It is possible to structure exams as "formative assessment"—to use exams to help students diagnose what they have learned and figure out what they need to change up to learn more. There are a few ways to employ exams as a learning experience. You could let students retake exams, or turn in their corrected answers to recoup points. If you emphasize group learning in your course, then group exams (in which a fraction of the grade comes from problems performed by groups) are becoming more popular. However, exams are typically a tool for summing up what students have learned at some kind of endpoint, and exams shouldn't be the principal way that students receive formative assessment, because by then, it's probably too late to help those who need it most.

While exams can be motivation for students to study, they are not necessarily an effective motivator. Exams tend to be the focal point of the semester. This is often a major source of stress, and the extrinsic motivator for studying. When students feel that the outcome of an exam isn't a predictable outcome of the time that they invest, it is difficult for them to translate this worry into effective studying throughout the semester. If we are going to put students through all of the anxiety of exams, we should make sure we have a good reason for making them go through the experience. It's difficult for instructors to provide intrinsic motivation to study, though that's what we need to aspire to. We are fooling ourselves if we think that the threat of a low grade on a high-stakes exam is enough incentive to study that we shouldn't work to create an intrinsic interest in studying.

Compared to other kinds of assessments, traditional exams have logistic difficulties. If students miss a high-stakes exam, administering makeup exams is often difficult, and it may not be fair to everybody in the course. If you are going to drop the lowest exam score in an effort to create flexibility for students, is this feasible for you if you only use two midterms and a final?

Traditional exams have a lot of evaluation biases baked into them. While students with documented learning disabilities may receive the appropriate accommodation, there are many students with undiagnosed disabilities (including severe test anxiety) who will underperform relative to their level of knowledge and skill. If you are using multiple-choice exams, biases that negatively impact students based on their socioeconomic and ethnic backgrounds are pervasive and problematic for even highly experienced exam developers. Asking yourself to write an equitable multiple-choice exam that manages to test the content that you wish to evaluate is a tall order.

When you are administering exams, what are you attempting to measure? Are you supposed to assess the cumulative amount of relevant information that a student has learned? Are you assessing their ability to perform certain tasks? Are grades supposed to measure the depth of understanding? In principle, the decision about how to use exams—or whether to use them—should be structured by what you want your grades to measure.

I recommend removing high-stakes exams from our classes, though in some environments this might be a hard sell because it goes against long-standing traditions. While you often theoretically have the academic freedom to make this choice, this may not be a wise choice with respect to departmental politics. If you end up doing what most folks are doing—having a few midterms and a final—then this chapter can help you steer through the process to make it as painless as possible for you and your students, removing the fear associated with the process so students can focus on learning while they study for their exams.

Setting the Difficulty Level on Exams

We shouldn't make exams that are harder than students will expect them to be. It's okay to be challenging, but if our exams challenge students in a way that you haven't done before the exam, this just blindsides them. If you want to challenge your students, you need to do this long before you administer an exam.

To land at a good average score, it's possible to make your exams extraordinarily difficult, with very low scores, and just curve the grade up by giving everybody a bunch of points so the average score is 75 or so. A student recently told me about their organic chemistry exam, on which they had gotten a 45% or so. I was trying to commiserate with them, then they told me that this score would probably end up being a B. This approach tends to make at least some of the students dispirited and doesn't help them learn. If an exam ends up being harder than you intended, giving a point boost to everybody can work. When you're choosing questions, it's not a good idea to try to find hard questions with the intention of getting the average down into the 70% range. Instead, ask the questions that you think best evaluate what the students have been expected to learn in the course.

When students say an exam is hard, this can mean one of several different things. An exam can be hard because students run into content they didn't even think would be on the exam. An exam can be hard because students might be asked to memorize a huge amount of material, stretching the capacity for information recall. An exam could be hard because there was a lot to do in a short period of time. Another way an exam could be hard is if the students are asked higher-level problems for content that they were taught at a lower level. For example, students were told they needed to know an equation, but were not prepared to apply it to real-world examples. Exams should challenge our students, but make sure that they're challenging in a way that's fair to students.

Three Kinds of Exams

In traditional exams, students are told what to study, then they come into the classroom, and then they see what is on the exam and take the test. Unless you say otherwise, students will probably assume the only materials they are allowed to have during the exam are a pen or pencil. (However, you do need to specify this in the exam instructions, in case you see someone consulting a phone or notes.)

When you give a traditional exam, be sure to let students know well in advance what will be allowed and what will be forbidden. Is a calculator okay, or is one necessary? If you want students to not have to fuss about memorizing some pieces of information, you could let them prepare a single index card or sheet of paper. Alternatively, you could tell your students that the exam will have a sheet of information available to them that will have specific information they might need (such as equations, biochemical pathways, bits of code, periodic table, or whatever).

Instead of a traditional exam, have you thought whether an open-book test would work for your course? This is an opportunity for you to evaluate your students on the higher levels of Bloom's Taxonomy. Especially in upper-division courses where you are working to get students to work on higher-level concepts, allowing students to use their books and/or notes will free them from having to memorize minutiae and give them freedom to work on broader concepts. The kinds of questions that you would ask on an open-book exam can be fundamentally different than the type you might ask on a closed-book exam. Regardless, the way that students prepare for open-book exams will be different. They won't be making flash cards to test how well they remember small bits of facts, and they won't be worrying about vocabulary and dates. Instead, they can focus on knowing how to solve problems and ensuring that they can explain, interpret, and apply concepts.

A third kind of exam that you can assign to your students is a take-home exam. For advanced classes where it can take some time for students to demonstrate a certain level of mastery, a take-home exam can do the job. It could be a set of essays, or a problem set, or one large problem, or a more creative assignment. One benefit of a take-home midterm exam is that you don't lose that instructional time. If you want guarantees that students do not discuss the exam with one another, then obviously a take-home exam is not a choice for you. I have done take-home exams with small upper-division classes and graduate classes that either involve problem-solving (such as biostatistics) or application and synthesis of theory (such as my behavioral ecology course). In those cases, it's worked well for me, but a small number of students were enrolled, and this would be hard to manage in a larger course.

Moving away from Midterms

Final exams are a well-established fixture, but more and more instructors are moving away from the model of a couple big midterm exams. Now that I've described different ways of running exams, I'd like to make the case that you can run your class more effectively without using midterms.

The main function of midterms is to give you numbers to calculate a grade. Secondarily, the first midterm often serves as a wake-up call to students that they need to change up their approach to the course. If you can find a way to give students feedback on their performance throughout the semester, and if you have enough assessments to give a final grade without midterms, how about just going without big midterms?

When I asked Matthew Venesky (assistant professor, Allegheny College) about his midterms, he explained: "I've moved from 3 exams plus a final exam, to 6–7 'quizzes' plus a cumulative final during final exam week. I initially made this move in the intro course that I teach, and I have since adopted it in my upper division course." This amounts to having a "quiz" once every few weeks during the semester. Venesky reports some major upsides to this approach: "I find that I can catch student misconceptions earlier in the course if I give a quiz sooner than 1/4 of the way through the course. I also find that the format of 6–7 quizzes allows me to use more questions that assess critical thinking via student writing responses, because the amount of content on any given quiz is less than a full-length exam."

On one hand, more frequent quizzes means more grading for you throughout the semester. On the other hand, this spreads the work throughout the semester instead of your having to grade midterms in a massive push. Since the educational research says that low-stakes frequent assessment is better for learning, perhaps it's worth giving it a shot. You're not subject to any rule that says you are required to administer midterm exams, are you?

Writing Good Exam Questions

Students are rightfully aggrieved when questions on exams are unclear or don't measure what they are supposed to measure. Writing bad test questions is also inefficient teaching, because this will generate student complaints that will subtract our attention from actual teaching.

What does a good question look like? Foremost, everybody will interpret the question the same way, and it cannot be interpreted to mean something that you did not intend. A good multiple-choice question has a single and unambiguous correct answer. A good exam question is one that none of your students can look at it and say, "I didn't think this was going to be on the exam." Good exam questions don't use vocabulary that some of your students might not be familiar with and don't make assumptions about student familiarity with historical figures, pop culture, literature, and so on. Good exam questions are not scaffolded so that a student has to get one question correct in order to answer a subsequent question.

Considering the ubiquity of multiple-choice exams, there is much research designed to understand what constitutes effective and fair multiple-choice questions. Here are some key findings. Every multiple-choice question needs a single and unambiguously correct answer. This means you need to stay away from "all of the above," "none of the above," or "A and C are correct." To make a difficult multiple-choice question, make sure that the incorrect answers are highly plausible but also demonstrably false. It's okay to have only three choices, and more than four is not advisable. If you wish for the test to enhance learning, low average scores are a drawback, but if it's too easy, this won't help either. A good question is one that nobody can credibly claim, "That was a trick question."

Sometimes I've written a question that I think is perfectly fine, but then I discover it has thrown students for a loop, even when it's clear that some of them have mastered that particular piece of content. In that case, I give everybody full credit for that question, essentially dropping it from the exam. I can tell a question is a problem if more than one student requests clarification or comes to me afterward with a good argument that their incorrect answer is actually correct.

Exams with Less Fear

Learning happens when we are focused on inquiry. We learn when we are interested, or excited, or curious. If a student is terrified of an exam, they'll spend time trying to be able to do well on the test, but it might not be possible to focus on learning the material. I've heard a lot of scientists equate "doing well on an exam" with "learned the main concepts that were taught in the course," but often these two things are not closely related to one another. If we can take fear and guesswork out of the exam for students, then they'll have a greater capacity to direct their studying effort on the material that we think matters. Here are many suggestions for steps that you might take to reduce text anxiety and make your class a more positive experience for your students.

It is entirely legitimate for a student to be told the basis of their evaluation. Students take a course and earn a grade. They should be made aware, as specifically as possible, the foundation for this grade before they do what it takes to earn it. The less they know about the basis of their evaluation, the less fair we are to our students. Specificity gives you control over the material that they will study. I have often heard professors say they are frustrated that students aren't focusing on studying the right material, or asking the right questions while studying. This is something that you can easily fix.

Well before the exam, you could give the students a written sheet of information that tells them specifically what to expect, which is explained in the next section. If you can lower the stakes for individual exams, this could be a tremendous help to students. The fewer points that can be associated with any single exam, the better. I try to make sure that no single exam counts for more than 20% of the total grade. If you can give enough exams to drop the score on the lowest one, this will free you and your students from having to do a makeup, or having to adjudicate excuses from students who are sick or have unexpected problems.

Another way to lower the stress of exams is to make sure they're not too long. As the instructor, and a super-duper expert in the topic, you're probably the least qualified to decide whether an exam is too long for students to finish without having to rush. Usually, we figure out whether or not our exams are too long by experience with administering exams that we have written. There will always be some small percentage of students lingering in the classroom hoping for inspiration to strike at the last moment, but if you have some students actively working to finish the exam right before the end of the assigned time, then your exam is too long. Jeramia Ory (associate professor, St. Louis College of Pharmacy) remarked that if multiple students aren't finished by the halfway mark, that's an indicator the test is too long. There's no need to make your test a race. Unless you're teaching a course in defusing time bombs or competitive cycling, your exams shouldn't put people who work slowly at a disadvantage. Some students needing more time because of a disability may provide you evidence requiring you to provide extra time. It also makes sense to accommodate everybody, including those with undiagnosed conditions, by making sure the exam is not too long.

Students will be more comfortable with exams if they have a better idea what to expect in terms of the format of the test. It's a good idea to let them know what the mix of question formats will be. If you have copies of earlier exams that you wish to make available to students, this could help. As I explain later in this chapter, there is no use in hanging on to old exams so you can reuse questions, because some of your students will have copies of old exams whether or not you have intentionally released them. I've sometimes made available copies of exams I've given in other courses, just so students could know how I format exams, so they will be less surprised.

Another way to make exams less stressful is to field questions from students in advance and make sure everybody has access to the answers that you provide.

Review Sheets

The most effective way to help students prepare for exams is to give them a review sheet, which tells them what they need to master. I think if we are going to be administering major exams to students, then the Fairness Principle requires that we provide students with a detailed set of criteria about what to expect on the exam.

When I was conducting interviews for this book, there was only one finding that really surprised me. When I asked, "When students ask you, 'What's going to be on the exam?' what is your reply?" a lot of the instructors were in favor of vague replies. One respondent would answer, "Everything we've learned here, and anything you've learned elsewhere that you think is relevant." Similar answers were "You'll find out!" and "Every topic we have covered will be represented on the exam in some way." A smaller fraction of my interviewees would tell their students, "See the study guide."

I think when we tell students, "Everything I said in class and everything in the assigned reading could be on the exam," this is a disastrous practice that disrespects the efforts of our students. This apparently is an unpopular opinion among many college science teachers, so I'll explain my thoughts. If we tell our students that everything is fair game, this will compel them to scramble to cover too many bases, by being familiar with everything, rather than working to truly learn a narrower scope of information. By making students attempt to infer what will be on the exam, you'll only help the students who are better at guessing what you think (which is a nonrandom subset of your students). When students don't have a discrete understanding of what will be on the exam and what won't be on the exam, the resulting anxiety keeps them from learning. You might not be the kind of person who would write an exam question about a small 30-second anecdote mentioned as an aside during lecture, but some instructors are actually like that. Experiences like those are traumatic for some students, and by giving them specific and discrete information about what will be on the test, we help them learn.

In each of your lessons, you hopefully have told your students about the specific learning outcomes that you've expected them to demonstrate. If you were adequately complete with these outcomes at the outset, your review sheet for each exam could be as simple as compiling all of these items together.

In chapter 1, I discussed the classification of different kinds of learning in Bloom's Taxonomy. It is most helpful if your review sheet was prepared in light of this taxonomy, paying particular attention to the verbs that you use for your expectation of student performance. For example, let's say you want students to know the first 50 elements on the periodic table. What would that look like on the exam? If you just tell your students "know the first 50 elements," that's rather vague. Instead, you can express your review sheet as a set of performance expectations. For example, "You should be able to fill in a periodic table with the names, symbols, and atomic number for all elements with an atomic number of 50 or less," or "Given an atomic number, you should be able to name an element and its symbol." That's a very simple example, and presumably you have more detailed learning outcomes from your students, but this might give you an idea. You probably have a good chance at finding some expected learning outcomes from some generous instructor who teaches a similar course and posts their material online.

I typically provide students with a comprehensive exam preparation sheet one or two weeks before the exam. Sometimes these items on the review sheet are very narrow, but other times they could be rather open-ended. But they are never intended to be vague. To keep myself honest when writing the test, I tell them that everything on the exam will correspond to at least one item on the review sheet. Moreover, I tell them that if any question on the exam isn't based on one of these review items, then I'll drop it from the exam. I am also tempted to hand out these review sheets at the beginning of the semester, but I tend to make too many course adjustments during the semester to make that a wise choice.

One strategy for the exam review sheet, which can motivate questions to think critically about course material, is to provide a list of actual exam questions well in advance. For example, you could tell your students that you will draw a certain fraction of the exam from practice questions at the back of the chapter in your textbook, perhaps with slight changes at your discretion. If you want to make sure that all of your students are capable of answering questions in the book, then you can simply let them know that this is exactly what they're going to be tested on!

The approach of giving students specific questions in advance also can work for other kinds of questions. You could give them a big vocabulary list that you will draw from. On some of my exams, I've put many short-response questions on the review sheet, and I tell students I'll pick a small number of them for the exam. This means my students will be spending their study time trying to find the best answer to those short-answer questions. If I pick those questions well, then I'm really helping my students learn. Likewise, if I expect the final will have an essay question, then I might give them five essay questions in advance and tell them that I will pick two for the final. I think this is much preferable to me telling them, "Learn all of this information, and the final will contain an essay about some of it." When students concretely know what might be on the final, then they will concretely study it.

Fielding Student Questions

You'll always have some students who will want to ask you very specific questions about what may or may not be on the exam—or they might be looking for you to amplify some topic and provide them with additional instruction about a topic that will be on the exam. You could handle these on an ad hoc basis as they come in, but there are a couple disadvantages to this approach. Responding to these questions with individualized and verbose replies creates a positive feedback loop. Instead of finding yourself fielding questions for a small number of students, it makes sense to provide this kind of input to the entire class. If a student asks a question that I think is downright useful for them, I make a point to share the answer with everybody.

There are a variety of ways that you can standardize your responses to student questions about exams, so that every student has the benefit of your feedback. You could post them to an announcement or a chat board in the LMS, or send an email to everybody enrolled in the class. There are online polling sites that you could use for this purpose, where anybody with the link can ask a question, and your responses will be visible to everybody. Alternatively, you could simply ask students with questions about the exam to hang on to them until class, and you could start or finish lessons by fielding these questions. I often schedule half of the class period immediately preceding an exam for discussing questions from students that have emerged from the review sheets. It often boils down to me saying, "I'm not going to reteach that entire lesson, but this is the nutshell version." In a lot of cases, it's students making sure that they're understanding the review sheet the way I understand it. Language is a slippery thing, after all, especially English, and some sentences can be read differently than you intended. This means when a student asks, "What did you mean?" I'll do my best to paraphrase, rephrase, and field questions.

In some institutions, it's standard practice for professors to hold entire review sessions outside regular class hours. (My experience suggests that this tends to happen on campuses where a high proportion of students live in on-campus housing.) In my opinion, holding review sessions at a time different than regularly scheduled class sessions goes against the Respect Principle. When your students sign up for your class, the university commits to teaching during scheduled class hours, and students then must schedule their lives around those hours. If you add an additional session outside those times, it's possible that a student might not be able to make it. They might be working, or be playing in an orchestra, or attending a different class, have sports practice, or perhaps have made commitments to family members. Even if you ask in class for students to let you know if a potential time slot is okay, any students with a conflicting schedule will not want to mention this and spoil the opportunity of a review session for the rest of the class. While you might imagine that the students in your course won't have any commitments at some particular time, they have their own lives that extend well beyond our courses. I think the only fair way to hold reviews is to do it in a scheduled class session, or in an asynchronous review online.

Grading Exams

If you ask an instructor to name the worst part of teaching, the most common reply is "grading." I concur. A pile of ungraded exams is no fun, but even worse is a week-old pile of ungraded exams. The sooner you tear off that Band-Aid, the better. When you put your exams on your syllabus at the start of the semester, it's a good idea to look at your calendar and block off time to grade those exams as well.

Before you dive into the pile of exams, I suggest taking an unmarked copy of the exam to use it as a grading rubric. (I cover grading rubrics more in chapter 5.) For every question on the exam, write down the criteria for earning points, and keep it handy. Then I suggest you go through your exams one page at a time. This means that you'll grade Page One on everybody's exam first. Then, before you dive into Page Two for everybody, take a quick flip at some of the first ones you graded, just to make sure that your criteria didn't drift as you went through the pile.

One thing you'll probably notice as you go through grading is that students might provide you with not-yet-complete answers that you did not anticipate when you built your rubric. When you decide how many points to assign to such answers, be sure to update the rubric, because if another exam has the same kind of error that you didn't anticipate (which is likely), then it's easy to stay consistent.

Another reason you want to update your rubric with all of the different kinds of wrong answers is that this lets you provide feedback to students in an efficient manner. It can take a lot of your time to correct student exams by giving them the correct answers. If there are long essays, your writing a few thoughts might be a good idea, if just to reassure the student that you gave their response the consideration it deserved. However, if a short answer has the wrong answer, you don't need to write in the correct one—just save it for when you hand the exams back. (Note that if the culture in your department is detailed feedback on every exam, then you may need to write feedback on exams.)

Handing Exams Back

Returning exams to students can be awkward. If you handle grades on the LMS, then some of this awkwardness can be handled in advance, by posting the scores before you hand the actual tests back to the students. If the exam was fully multiple choice, then you have no need to return exams back to students. However, if the exam involves short essays and problems that were solved, especially involving partial credit, it's good practice to return these assignments to your students.

If you did a solid job taking notes on your blank exam that you co-opted as a grading rubric, then you should be well equipped to take 10 or 15 minutes out to walk your class through the exam and describe how different responses were graded. This is a lot more efficient than writing remarks on every single exam to explain how points were assigned, and as far as I can tell, students haven't seemed to be bothered by not getting more detail. Alternatively, you could post a key to the exam. It's my own preference to verbally walk students through the exam instead of providing a key. This way, every student in the room gets to hear about correct answers on the exam, whereas I don't have the assurance of that by posting or distributing a key. This also is an opportunity for students to ask follow-up questions that might not be addressed in a key. I think either way is fine, or you could discuss the key in class and then share it afterward. In the next chapter, I discuss how to respond to situations when students ask you to reevaluate your exam for more points.

When You Teach the Same Course Many Times

You could be teaching in the biggest university or in the tiniest liberal arts college, but if you are teaching the same class more than once, you're going to be facing the reality that copies of your old exams are going to get out there. If you're teaching the class over and over again, this might make it hard to run exams that have a level playing field for all students. I might remind you at this time that if you adopt a variety of alternatives to traditional exams, you might not have to deal with these problems.

Long before the days of the internet, students often maintained "test banks" in fraternities, sororities, student clubs, and elsewhere. Nowadays, I imagine such test banks still exist, but there are all kinds of online repositories. The technological landscape evolves quickly, but one constant is that there are always folks out there working to game the system. I don't think it's constructive to go to extraordinary efforts to prevent your course from being gamed (because you're never going to outplay the players, and this is an evolutionary arms race with the faculty consistently lagging behind). Nonetheless, it's unwise to wholly disregard the common steps that students might take to gain an unfair advantage on your tests. This is one of those areas where I think it's wise to consult with seasoned instructors to find out how they approach this situation on your campus, but I'll describe some common strategies to deal with this challenge.

A common tactic, particularly in large courses, is to prevent students from keeping a copy of their exams. If the exam is all multiple choice, then students can receive a final exam after their cell phone is put away and then have to turn in their copy of the exam to be able to turn in their Scantron form. In theory, this would prevent any copies of the exam from circulating. In practice, this doesn't work that well. Ultimately, if you try to prevent people from keeping exams, copies will get out one way or another.

Instead of trying to lock down access to test questions, you could open them up so they are available to everybody. One way to do this is to make copies of previous exams available to the entire class, by posting them to the LMS. On a longer time scale, once you have accumulated a bank of test questions that you think are high quality, you could make the entire test bank available to all of the students. After all, if students are capable of answering all of these test questions, then you will have been successful at teaching, right? (If you do have a bank of questions like this on the LMS, you could administer them as a series of quizzes every week or two. You could even allow students to retake the quizzes multiple times, using randomly selected questions, until they earn the score they want. A number of large universities do this in their introductory courses.)

Another way to deal with the possibility of students knowing what will be on the test is to make sure that students will know what will be on the test! If your course is small enough that the exam doesn't have to be multiple choice, then you could simply give your students a bunch of questions and problems in advance, and then draw up your test as a subset of those questions. In this case, it's likely that you'll have students sending you emails and visiting you in your office, asking you questions about these items that might appear on the exam. I see this as a huge win! When interacting with these students, I think it's important to not give any individual privileged information that would give this student a relative advantage over other students. An appropriate way to handle this is to make sure that any replies to written queries are made available to the whole class (by email or via the LMS), and after you have a session of office hours with students, you could recapitulate the conversation with all students in print or in the next class meeting. Sometimes when a student asks me a good question in my office about a specific topic, and I realize it would be unfair to answer that particular question just for them, I tell them that I'll start out the next class by answering that question for everybody. You can also plan in advance to put time in the course schedule to review student questions while everybody is in the classroom.

There are schools of grading approaches called "criteria-based grading," such as specifications grading, that can help you deal with this test question problem. This is a whole topic unto itself that can't fit into this book. I'll just mention that if your main learning objectives require students to demonstrate proficiency at performing certain tasks, then your exams can simply be an opportunity for students to demonstrate that they can perform those tasks, and students can have other avenues to demonstrate mastery of those tasks. This approach sounds particularly apt for classes in general chemistry, math, and physics.

To some extent, your approach to the reusability of test questions can be adjusted by reconsidering how and why you use tests. It's possible that you could use tests only to evaluate basic recognition and understanding (lower-level processes in Bloom's Taxonomy). If that's the case, you can just reuse the same items over and over. If you want to evaluate how well students can demonstrate more sophisticated processing of information, then perhaps it's easier to save this for upper-division courses where class sizes are small enough that you can use alternative approaches.

Exams as a Learning Experience

Exams are typically used as summative assessments. However, they also can be designed as genuine learning experiences, by making the exam a two-stage process. The first part of the exam is worth credit, and the second stage involves an opportunity to recover points lost in the first stage.

You could let students correct the errors on their exam and turn them in to you for regrading. (Obviously, this will not work if you distributed an answer key to students after the exam.) In these circumstances, it is common to give students credit for this work at a fraction of the original value. For example, if a student took an exam and earned a 70%, then when you return it to the students, they have the capacity to earn back 50% of the points they did not initially earn, so that an exam that is fully correct would garner 85%. Some instructors do not initially plan to allow students to correct exams for additional points, but if the exam was more difficult than they had intended (either because of a lack of student preparation or a miscalculation in exam design), then they can decide on the fly to give students this option.

An alternative approach to correcting exams is to allow students to re-take an exam in its entirety. You could simply grade both exams and enter the higher grade in the gradebook. This is more frequently employed in classes that involve a substantial amount of mathematical problem-solving.

Another way to convert exams into a formative assessment tool is to run cooperative exams. Cooperative exams are run in two stages. First, students are required to do independent work on some exam questions. After those answers are turned in, students then work in groups. These groups could work on the same questions as the independent exam, or on some new questions. The grade for each student is composed of both the independent and the group score, though it's typical for the independent score to compose at least 75% of the final grade. If you are having your students work in well-established groups throughout the course, then cooperative exams might be a good fit.

7

Common Problems

Catching Problems Early

If you experience major hiccups once in a while, you are not alone. Teaching is hard, and if you are trying new things, sometimes they don't work out. This is normal. Difficulties turn into actual problems when students become downright disgruntled by the choices that you've made. Even a small number of disaffected students can taint the learning environment for an entire course of students. It gets hard to focus on the subject matter when there is an air of dissatisfaction in the community. It's possible that you actually have botched the job and some students are right to be upset. It's also possible that you didn't do anything wrong and the students are out of line. Regardless, if there's a perception in your class that things aren't going well, this makes it hard for everybody to learn (and might also be a professional liability for you), so it's a situation that has to be addressed, regardless of culpability. Because even the appearance of things not being right keeps you from being an effective instructor.

My experience is that by the time I diagnose a problem, it's already grown to the point where it's hard to fix the situation. I feel like I've screwed up enough in the past that I now tend to avoid big problems. Regardless, some problems will emerge even if you try your best to avoid them. In addition to the Respect Principle, I have three specific steps to recommend that can help you avoid major problems throughout the semester.

First, try to identify an instructor you respect, who would enjoy serving as your mentor. This person doesn't even have to be in your own specialty or department. They just need to share some key values with you and show an interest in helping you grow.

Second, once you get a whiff to suggest that things might not be going smoothly, I recommend soliciting anonymous feedback from the whole course. You can just come up with a single sheet of paper that has a set of general and specific questions about areas that concern you. Ask students what they like and don't like, changes they might want to see, and how the course has and hasn't met their expectations. If you administer this at the start of class and give everybody 5–10 minutes to complete it, you'll probably get more useful responses than if you pass the form out near the end of class and ask people to turn it in on their way out the door. Doing a midsemester evaluation is a good practice even if you don't think you're having major problems.

Finally, it's always worthwhile to check out your campus Center for Teaching and Learning (though it might go by some other name). There are experts in college teaching on your own campus who are there to help you improve your craft, open to all instructors. In addition, many universities have programs that provide specialized training and teacher certification for graduate students. These offices provide not only seminars and resources for you to access but also there will be teaching experts who can spend time with you on an individualized basis. Don't be reluctant to take advantage of this resource! Gaining professional development as a teacher is not typically seen as remediation for the instructor, but instead as an extra step taken to improve student success.

Grade Change Requests

Every instructor faces a situation where a student will approach them and ask for a higher grade. This happens either when the student thinks that we made an error, or perhaps when we didn't make an error but the student feels like they still need or want more points. It's normal to immediately feel defensive when a student approaches us with a grade appeal. Following the Respect Principle, we need to swallow that defensiveness and hear the student out. If their request has no merit, then we can make that decision after following a clear process. Because any student can just choose to come up to you and ask for more points, part of being fair to all students is to have a clearly defined and transparent process to deal with grade appeals. We should try to lower the bar for students to approach us when they genuinely have been shortchanged on points accidentally, but we also don't want to overhumor any student who is simply looking to milk out more points than they earned.

My broad philosophy about grade change requests is to avoid conflict while working to make sure that I got it right and that all students are being treated fairly, including those who are not asking for a change in their grade. While a student might be more concerned about their grade than I am, I am more concerned that they remain engaged with the course. An adverse interaction about a small number of points can taint the whole well. Of course, I would never go so far as to give a student unearned points because I thought it would make them happy, but I do pay attention to their feelings, as a fellow human being.

I tell my students that if they've found a basic arithmetic error, they should bring it to me right away, and I'll take note of it and fix my grading sheet. However, if a student comes up to me thinking that I made the other kind of mistake—a judgment error in assigning points—I'm not going to stand there in the classroom after the exam and regrade the test for students. What I ask them to do is to bring the test back to my next class (or my office hours) with a very short note explaining where the mistake was made. This can cut back on frivolous requests for unearned points, but it also provides an avenue for students who genuinely were subject to an error on your part to clear up the matter with a minimum of confrontation on their part.

Sometimes a student can be particularly persistent about a very small number of points, when they feel they got shortchanged on a question that they thought was unclear or unfair. Sometimes it doesn't help to show that, number-wise, there would be no practical impact of the grade change. To these students, receiving points for that question is a matter of principle, and clearly they feel strongly about it if they're bringing it up even though it doesn't have a real impact on their grade. If I agree with the student that the question wasn't as clear or as fair as it could have been, then I'm inclined to give every student in the whole class the extra points, effectively dropping the question from the quiz or exam. If I think the student's concern is not valid, then it's my job to work with the student to discuss the issue, so we can understand the nature of the misunderstanding. Sometimes you can get a reality check by running these items by a colleague whose judgment you trust.

Student Privacy

The expectations for student privacy are evolving rapidly, though standard practices may vary dramatically among institutions. Some universities have practices that are not up to snuff, while others robustly work to protect student rights. At a minimum, if you are teaching in the United States, you should be aware of the protections ensured by FERPA, the Family Educational Rights and Privacy Act of 1974. In short, FERPA guidelines require that you not disclose private educational records to anybody who doesn't have prior approval by the student. If you are not teaching in the United States, you should familiarize yourself with similar laws in your own country.

I think the protections provided by FERPA are all entirely reasonable. If you stick with the letter of the law, then you'll be doing just fine in respecting student privacy. Likewise, if you end up pushing the boundaries of what is allowable under FERPA protections, then you're also going to end up stretching the boundaries of a respectful relationship with your students. Even if you're not bound by US law, it's still a good idea to follow these simple guidelines. Essentially, the law says you are not allowed to disclose any information about a student's educational records in a way that can be identified with the student. Your institution will have a website to describe its FERPA policies, and this is worth a browse. I'll walk you through some common situations that have FERPA issues.

Perhaps the greatest potential for a violation of student privacy occurs when we hand graded work back to students. While I suppose there are some students who don't mind when other students are able to see their scores, you definitely will have some students who would rather their scores be private. It's good practice to leave it up to them to decide if and with whom they will share their scores. It is a violation of privacy to leave graded assignments out for students to pick up (either on a table in your classroom, or in a box somewhere); that means other students can fish through the papers and notice the scores of other students. A simple way to deal with this is to hand papers back to students yourself individually. (This also is a great way to learn the names of students!) While this potentially could take up a lot of instructional time, if you hand papers back while students are doing a group-based activity, then this won't be so disruptive and won't take time away from the lesson. If you wish to array papers in alphabetical order for students to pick up as they walk past, this is typically fine as long as the grade is not visible on the front page and the pickup is under your observation. If you really need to have students pick up assignments when you are not present, then the only proper way to do this is to arrange for this through an authorized third party, such as a departmental administrative assistant, though I think few of us have staff who have this kind of task in their job description.

Posting scores in public with any individually identifiable information is also a violation of policy. I remember when I was an undergrad in college, they just posted our math placement test scores on the wall next to our names. That kind of thing cannot fly nowadays. Some instructors deal with this issue by using student ID numbers instead of names. Keep in mind that in at least some institutions, this is still interpreted to be a FERPA violation, because this is an individual identifier that can be traced back to a student. One upside of using the LMS is its design for student privacy. Every assignment that you handle online is obviously one that you don't have to deal with on paper.

Depending on where you are teaching, you might need to be prepared in advance about how you'll respond to being contacted by a parent or other relative of one of your students. All adult students—in the United States, that means all students over the age of 18—control access to their educational records. Even if a student's parents are fully paying for their college education, these parents do not have a right to be informed of the educational records of their children. Not only are you forbidden from sharing scores with these parents, but any discussion about their academic performance is prohibited. If a parent calls you up, and you discuss how this student is doing in your class with them, then you've just violated the law. In these circumstances, the best way for you to handle this is to simply tell parents that you are not allowed to discuss this matter, that you are forbidden by federal law.

Of course, students can waive their FERPA rights. It is becoming increasingly common, especially in private institutions, for students to receive a parental FERPA waiver to fill out when they first enroll. In some institutions, apparently the default is for students to fill these forms out even before they step into a classroom! While it's necessary to assume at the outset that your students have not waived their FERPA rights, once you have written evidence of a FERPA waiver, then it is possible for you to have a discussion with a parent about academic performance. Even if a student has waived their FERPA rights to share their academic records with their parents, this does not mean they have empowered their parents to conduct business on their behalf. The FERPA waiver means the parents can be informed about what is going on in the class.

What might a parent of a full-grown college student want to talk to you about? Not surprisingly, it might be about appealing a grade that you've already assigned. It also could be about scheduling matters. For example, if an exam in your class conflicts with a family wedding, a parent might contact you to try to arrange for an alternative way to take the test. It's also possible that the parent may be aware that the student is not earning the grade in the course that the parent wants them to earn, and they are merely seeking your feedback about the kind of support a student may require to improve.

If you do get contacted by parents or other family members, it's because they're concerned. And perhaps because they're probably accustomed to meddling into the affairs of their adult children. Emotions are presumably running high by the time they're contacting you, and the prescription for such a circumstance is to tread lightly and wisely. Before engaging with family members of a student with a FERPA waiver, it would be wise to confer with two parties. Foremost, consult with the student about what they wish for you to disclose to their family members. Just because a student has signed a waiver to give other people access to their records, that doesn't grant the authorized parties free license to probe more deeply into the affairs of your student. Second, I think it's wise to consult with a trusted colleague who has experience on campus dealing with family members who have a FERPA waiver. The culture and expectations of every campus are different, and the further your own actions stray from institutional norms, the more resistance you'll meet. Be sure to protect the interests of your student and follow your conscience, and step into these conversations with an understanding of where the other party wants to go.

Keep in mind that a signed FERPA waiver doesn't mean that parents are entitled to know everything. The waiver just means that you aren't breaking the law by disclosing information because you have been authorized by the student. It's perfectly legal to not disclose information, even if a FERPA waiver is signed. You're not testifying under oath and compelled to answer questions that you don't want to answer. It's typically in the best interest of students for them to work with you on their grades, studying, schedule, and so on. Under normal circumstances, these matters are just handled directly with students, so understanding how the circumstances differ from the norm can help you navigate the interaction. Of course, if a student has a request that you cannot honor with full respect and fairness, then it would make sense that you can't honor such a request after parental intervention. If this conversation appears to be purely about parents doing a power play to change your mind, then this is just the kind of situation where departmental chairs are expected to run interference on behalf of the faculty. If someone needs to be mollified, this might be above your pay grade. However, if you think bringing in your chair would not be favorable for you, then obviously it would be best to handle this yourself. Again, I'd like to emphasize that this is one of those situations where the advice of experienced senior faculty in the institution can be useful.

Disruptive or Threatening Behavior by Students

Student disruptions to class are rare, but it is helpful to be mentally prepared to know how you will take action if such an event occurs while you are teaching. If a student's behavior has escalated to the point that normal operations in the class have halted, then you can firmly and calmly instruct the student to leave the room and tell them you will be available to talk to them after class is over if they wish.

If this student does not leave at your request, this is where you have to exercise your best judgment to minimize the risk of harm to all parties involved. It is likely that highly inappropriate conduct by a student in class is a manifestation of mental illness, and these are circumstances where they will need support and treatment, rather than an episode with law enforcement. While university police and campus security personnel are trained to deescalate situations, bringing them into the situation is an escalation itself and might put individuals in the classroom at greater physical risk. If your disruptive student is unarmed and your campus security are armed, then that's a phone call you might not want to make on the spot.

If you receive any kind of communication from a student that has you concerned about the possibility that this student might harm themselves or someone else, share this with your chair and, if the situation is urgent, immediately with campus safety/police. If a student is expressing intolerant views or bigoted remarks that rise to the level of hate speech, then this is a matter to discuss with your chair and use the proper avenues for reporting.

If you are in the United States, then like myself you must be acutely aware of the possibility that an active shooter may appear on campus and target you and your students while you are in the classroom. I don't have any particular wisdom to share other than what has probably already occurred to you. When you start teaching in a new space, it's good to note routes of egress (which is important for emergencies such as earthquakes or fires, as well). Note whether the doors can be secured from within and how you might stop visibility from the outside. I would not be surprised if your campus has posted directions in every classroom about what to do in the event of a campus alert about a lockdown or "run, hide, fight" protocol. If your campus has an automated emergency system, it's a good idea to make sure your mobile phone is registered in the system.

Disrespect for Your Authority

At the beginning of the book, I discussed how to step into a role of authority and how to manage a classroom. If things have gone south, here are some thoughts about recovering from the situation.

It helps to be realistic. Once you've lost a course—that means some students just don't have respect for you and what you're doing—you're not going to win back that respect before the end of the semester. At this point, instead of planning a massive comeback, it's better to plot for a recovery of what you have left. To start this salvage operation, you need to figure out the underlying circumstances and come up with a damage report.

Some people are more likely to be targets of disrespectful behavior than others: women, members of ethnic minorities, professors who appear to be young, and instructors with other marginalized identities known to students. If you are not new on campus, then students enter your classroom with a full set of expectations generated by your reputation from prior semesters. Once you have a reputation with students, that's your reputation, and it's hard work to move that needle. If you are having problems with students, it's useful to figure out how much of the situation has been triggered by your identity, or some of your own actions. Getting a clear read on this situation is hard when you're in the middle of it.

I think disrespect issues can be lumped into one of two general categories: The first is when a small number of students have some kind of beef with you, and most of the students in the class are just fine. The second kind is where you feel like you have just lost the entire class—you just don't have a good relationship with the class and they just don't have confidence in you. I think diagnosing this distinction is important because if you've lost the entire class, that means you need to figure out what you might have done differently, to prevent this from happening again.

One particular kind of challenge hits graduate teaching assistants and brand-new faculty, when students don't accept their instructor as an authority. If you aren't much older than your students, it might be tempting to try to remove barriers by interacting with the students as your peers. This is particularly difficult if you just finished your undergraduate degree, and now you're running a lab section of undergraduates! Are you actually in charge of this lab, and do you really have to assume the role of the authority figure? Well, you have to. If you buddy up with your students, it will end badly, because in the end, you are the authority, and you do have say over policies and grades. An artificial dissolution of the instructor/student boundary will end up with conflict when students try to cross that line. What can you do, if you're already buddying around with your students, and now you need to act like the person in charge? Unfortunately, the horse is out of the barn. To try to fix things, what you can do is dress more formally, avoid talking about things not directly related to the course, avoid saying anything that would minimize your responsibility for what happens in class, and, if appropriate, review the section about TAing a lab section in chapter 4.

If your students are sour on you, they have their reasons for being sour, which may or may not be legitimate. You can't make things perfectly sweet, but you might be able to make the semester less tangy by making a clear effort to systematically assess what students like and dislike about the course, and then making an honest effort to make changes to meet the needs of the students. Especially when I think a class isn't on track, I make a point of administering a midsemester evaluation. I type up a one-page form that has a combination of general questions (like "What do you like about the class?" and "What aspects of the class are not helpful?") and specific questions about what I think might be problem areas (such as "Do you think the quizzes have not enough or too many multiple-choice questions?" and "Do you think the reading assignments are a good length and useful, or not?"). After reading the responses, if you find a common set of grievances, your class might respond positively if you make changes to improve their experience. Keep in mind that if you have a policy that students actively dislike, this policy is disrupting the learning environment. I'm not saying you need to cater to your students' whims or let them set your policies, but I'm pointing out that if you haven't won your students' buy-in, it'll be a lot harder for them to learn and more difficult for you to teach.

Once you have a poor relationship with your students, it's nearly impossible to fix the situation over a short time frame. The good news is that, no matter how you try to repair a bad situation, it will be over soon enough. You can always learn lessons from a bad semester and use these lessons as you move forward.

Title IX and Sexual Misconduct

Given the perennial epidemic of sexual assault on university campuses (and beyond), it is probable that students you are teaching have experienced trauma at some point during the course. As you navigate through the semester on a daily basis, it's helpful to keep this reality in mind. If a student is struggling academically, they might be having issues tied to experiences outside the classroom that they will not volunteer to you. It's not our job to serve as counselors or investigators, but you should be aware of the steps to take if a student does end up asking you for help. It is possible that survivors of assault might decide that you are a trustworthy person who can provide support and disclose information about their assault to you, or you might be notified secondhand of information shared by a student. In case this happens, you should be aware of university policies and how to respond with appropriate support.

In the United States, policies regarding sexual assault and sexual harassment in universities are subject to Title IX. The most relevant piece of Title IX that you need to be aware of is whether your institution classifies you as a "responsible employee." Under Title IX, all "responsible employees" are mandatory reporters, which means that you are required to report to your campus Title IX office every report that you hear of or witness involving sexual misconduct. In many universities, all instructional faculty are "responsible employees," but this is not universally true. If you're a graduate teaching assistant, you may or may not be a mandatory reporter, depending on how you are employed and how your university has developed its policies. But if you are in a situation where you might expect students to volunteer information about a sexual assault (which is far more common for women faculty), you'll need to know what to tell the student about your Title IX obligations.

Reporting to the Title IX office does not necessarily mean that a university will be launching an investigation. The Title IX office is obligated to contact the student to inform them about the report and provide them information about their options and support services. The student should not be expected or required to file a complaint. You should also be forewarned that many campus Title IX offices work to protect the interests of the institution over protecting survivors of sexual assault.

If you are a mandatory reporter, then you should probably let students know this status as they start to bring up an incident with you. You can also tell students that they are completely free to discuss "hypothetical" events with you in as much detail as they wish and that this does not trigger your duties as a mandatory reporter. Please be sure to look up the resources that are available on campus to support survivors, so that if a student needs support, you know how to put them in contact with the appropriate parties, such as a confidential advocate or the women's center. Also keep in mind that many campuses seek to prevent students from filing charges with the police, as it is common to protect the interests of the institution over the interest of individual students, so seeking support from off campus may also be appropriate.

When a Student Monopolizes Discussion

Once in a while, you'll get a student who thinks very highly of their own thoughts. So much so that they'll share them with the class far more often than everybody else in the room would like. This behavior harms the educational environment for other students, and instead of just tolerating it for the whole semester, it's better to deal with it. If this student is annoying you with their constant contributions to class, then they are probably even more annoying to other students in the class. Of course, the Respect Principle applies to all students, even the talkative ones who disrupt class, and it's wise to avoid a snappy or passive-aggressive remark at their expense. There are a variety of ways of approaching this, depending on your personality, the nature of the course, and the students you're dealing with.

I think the constant contributors can be classified into three categories. The first are students who are just genuinely excited and curious, and when they ask questions, they really want to know the answer. The second category are the know-it-alls, who feel compelled to demonstrate their brilliance, but really this is just caused by the Dunning-Kruger effect. The students in the first category are less likely to take offense when you take measures to limit their contributions, while the smarty-pants ones might resent that you're trying to quiet them down. The third category are students who are neurodivergent (for example, diagnosed with Asperger's). I recommend taking particular care to be respectful of students who are interacting in good faith and may not be aware that their participation may not be supporting the learning of everybody in the class. Regardless of the reason why a student may be monopolizing discussion, there is little difference in how you handle the situation.

You might consider talking to the student after class ends, after everybody else filters away, and ask them to chime in less often. They are probably unaware that their contributions are too frequent or too strident, and while they might not be overjoyed to hear this from you, they presumably would appreciate your discretion. It could be useful to couch this conversation in the context of metacognition, that it's important for all students in the class to be aware that they have their own opportunity to discover information on their own.

Of course, you might want to run your class so students can spontaneously ask questions or make remarks. One way to deal with this is to establish a "rule of three" for everybody in the class—nobody is allowed to chime in a second time until three other people have contributed as well. You could make that number even bigger, though it's best to pitch this in a way that your exuberant student doesn't feel singled out (even if that is your intention). If your classes involve whole-class discussions, this might be one of the norms to establish at the beginning of the semester.

Another option is to wholly change up how you run the classroom, so no student is expected to chime in by speaking or raising their hands. There are some suggestions about how to go about this in chapter 4. If this doesn't quiet the overenthusiastic student, they might need a gentle reminder that classmates need to participate too.

Overentitlement

The Respect Principle becomes more of a challenge when we interact with students who do not reciprocate the respect that we show to them. Some students honestly are not aware of our cultural norms for the kinds of accommodations that a student may expect from their instructors. Some of our students will not be respectful of our time. When this happens, it's just a part of the job. How should we handle interactions with students who act as if they are entitled to special treatment? It's simple: we must act professionally. Just like other potential problems, the best course of action is to create an environment that reduces the incidence of adverse events and to deal with these incidents properly and efficiently when they crop up.

How can we minimize the times when overentitled students ask for things from you that you can't or shouldn't do for them? In general, I think these requests fall into two categories: time or points. Students who have time problems want to turn in an assignment late, or take a makeup quiz or exam, or have an excused absence from class. These kinds of requests usually crop up when you teach with no-tolerance policies, such as not accepting late assignments, not offering makeups, and not dropping the lowest score. When your policy is "no exceptions," that is precisely the situation when overentitled students will be seeking exceptions from you. As suggested in chapter 2, when you build flexibility into your grading scheme (by dropping one or two of the lowest scores, and by having a light policy on late assignments), this isn't just being more fair to all students, it will also cut down on inappropriate requests.

Obviously, there's less you can do when students want unearned points or a higher grade. Extra credit is discussed in chapter 5, though I don't recommend giving a student an opportunity to earn extra credit just because they asked for it. I'd like to emphasize that all students should have equal opportunity to earn any extra credit that you make available. When students come to you asking for more points, they are often in an emotionally sensitive position, especially as the term comes to a close. In chapter 6, I provided suggestions on handling students who are asking for additional points on exams because they feel that they earned more points than you assigned to their response. How can you turn away a student who wants to argue with you about how many points they should have earned on a test or a quiz? While some students might be trying to milk out unearned points, a more parsimonious conclusion is that a complaining student feels that their professor shortchanged them of points, either accidentally or on purpose. It's worth your while to spend time with the student to explain the question and how points were awarded. There is nothing to be gained by debating a student about the merits of your grading decision, but explaining the science of the question itself can help them understand. Keep in mind that while these decisions might seem completely transparent and unambiguous to you, the student might be reading in shades of gray, as they do not have the same mastery of the subject matter as you do.

Student overentitlement can grow into a real problem that distracts from our teaching and spoils the learning environment in the classroom. However, we have the power to prevent this from happening, and we can do this by changing our own perspective. A lot of professors will respond to student overentitlement with shaming behaviors. This is often hidden from students, but in some work environments, it's common for faculty to mock students as rude, inappropriate, or spoiled. For those who mock students in private, I imagine it's hard to switch gears when you step into the classroom and must behave professionally for students. How do you see yourself as a professor? If you conceive of yourself as boss, parent, or commanding officer, then I can understand how you can take overentitlement as a personal offense. On the other hand, if you are a coach or personal trainer for your students, then when students have unreasonable requests, as a coach you can just say no in a firm but friendly manner, and give them the room to practice for the next sporting competition.

Underentitlement

While overentitled students get all of the press and are known problems for the instructor, a serious challenge for students lurks under the radar, and it's very difficult for us to detect. While you might have overentitled students lying to you about family funerals, you also definitely have students who are going through real difficulties that you are unaware of. They need support, too, but they're not asking for it. Underentitled students feel deterred from speaking up for themselves and end up getting treated unfairly compared to other students in the class. Underentitlement is a pervasive problem in academia, but one that we rarely talk about. Let me offer some examples to explain how underentitlement manifests in the classroom.

I was working closely with a student under my supervision in my research lab. A year earlier, this student had gotten a B in a class that I was teaching, because they had not turned in a major assignment due at the end of the semester, which was worth 10% of the total grade. If it weren't for that error, they definitely would have earned an A in the course. At one point, I asked, "What was up with that report that you didn't turn in?" They told me about a convergence of financial, health, and legal crises in their family, beyond their control, but still causing a major impact. It was the kind of situation that, if the dean of students were informed, the student would benefit from intervention in the form of an Incomplete or some other accommodation. I asked them why they didn't say anything about it at the time. They said, "I just didn't want to inconvenience you." I believed them, and I also think this student didn't want to enter an awkward conversation about their family with their professor. Now that I had gotten to know them well, we had finally had this conversation. But otherwise, I never would have known that the reason this person got a lower grade wasn't about a lack of personal responsibility but because of an excess of personal responsibility.

On another occasion, after I returned a set of exams to students, one student brought their exam to me the next class period, because I had made a very obvious mistake and didn't award the full points for a question that they got correct. I corrected the problem, but then I realized I was pretty sure that I made this error on every single exam I had graded! After I mentioned this to the class, I learned that a few other students had detected the same error, but those students were reluctant to come forward to mention the problem. This was a lesson to me in professorial power, how students might not be prepared to challenge my authority, even when I am in the wrong and I would like them to do so.

By communicating to students that we are accessible and open to discussion, we can try to break down barriers to help underentitled students advocate for themselves. However, just encouraging our students to tell us when they need accommodations or support won't bridge this gap, and it also may open us to exploitation. As described in chapter 2, the best way to address the problem of underentitled students is to adopt policies that give leeway to all students without anybody having to ask. This doesn't create an even playing field, but it makes it less slanted.

It's possible that your students are facing personal challenges that go beyond your capacity to accommodate them. Perhaps they were in the hospital with a broken femur in traction for nine days, or had to take time away after an incident of domestic violence, or had a very ill member of the family. If you can't deal with this seamlessly, then this kind of situation is genuinely above your pay grade. Students with extreme circumstances should make a point to visit the dean of students' office, who should be able to coordinate with faculty on their behalf. If any student has a situation that sounds dire enough that it will genuinely interrupt their studies, this is where to refer them.

Low Attendance

If fewer students are showing up than you expect, I suggest thinking of this not as a problem in itself but instead as a symptom of a problem. The bottom line is that students aren't showing up because they think they have little to gain from attending, or they have higher priorities.

One question that you'll need to ask yourself: Is missing class actually a problem for your students? Will it affect their learning or their grade, and if so, how? If your lessons are all available online, and if students don't need to be in class every day to earn points, then I don't think low attendance is, inherently, a problem. But if students are missing out on irreplaceable educational experiences or points, then this isn't a good situation, and it makes sense to be concerned. This is just the kind of situation that is well suited for taking to the campus Center for Teaching and Learning, or for consulting with an experienced mentor.

Some of the most instinctive fixes to attendance problems can be counterproductive. If you choose to take attendance, or give quizzes without warning, then this might get more students into the door, but it might also produce resentment, which will get in the way of learning. I think using points from clicker questions throughout the semester is a perfectly fine way to assign points to students, as long as students who miss a few classes don't suffer adverse consequences.

Once attendance is chronically low, it's hard to imagine what changes you can make to get students back in the door on a regular basis. Once established, patterns of behavior like this are difficult to change. Instead of focusing on what you can do to compel students back into class in the middle of the semester, I suggest thinking about how you might finish the semester out to provide the greatest experience for those students who do show up in the room. Then, if you're teaching in future semesters, you can consider changes to your course that will provide students positive reasons to show up in class.

What kinds of classes bring students to attend? What can you do other than assign points to students who attend? Be sure that you are giving students an in-class experience that they can't get outside class. For example, if your lesson consists of a highly detailed slide show that recapitulates the full lesson in text, then it's only natural for students to think they can miss attending lecture and get caught up by reviewing the slides at a later time. If your lesson relies heavily on resources provided by the textbook publisher, then students might conclude that they can miss class and catch up by reading the textbook. On the other hand, if your class has active learning built in—and the lessons learned from these activities are important for assignments, quizzes, and exams—then students who attend class will be rewarded with learning opportunities that they can't get from the LMS or their textbook.

Teaching Science Dismissers

No matter what science discipline you're teaching, it is likely that some fraction of students in your class do not accept foundational facts in your discipline. You might have students who dispute the age of the planet or the origin of our own species. Some might think that condensation trails contain chemicals that affect human behavior, or that it is not thermodynamically possible for the heat generated from the combustion of jet fuel to melt steel. Some might think that all transgenic foods are inherently unsafe to eat. They might think that structures built by earlier civilizations were not built by people but by organisms from a different planet. They might not think that global climate change is being driven by the industrial release of carbon into the atmosphere.

How can we teach students who don't even believe the basic facts that underlie our disciplines? A famous evolutionary biologist once said, "Nothing in biology makes sense except in the light of evolution," so it follows that it might be very difficult to make biology sensible to students who don't accept the fact of evolution. Regardless of your discipline, teaching content that contradicts what a student has previously learned is a professional challenge. If we are aware that some of the students in the class are dismissing scientific knowledge based on poor evidence or religious beliefs, this is something that we should not ignore. Our job is to teach all of our students, by giving them the greatest opportunity to learn. We create an environment for genuine learning by meeting our students where they are.

When students are actively dismissing the validity of a scientific field that is the basis of our careers, we might have to dig deep to root ourselves in the Respect Principle. Yes, effective teaching requires that we respect them as individuals, including all of the ones who might be young-earth-creationist global-warming-dismissing anti-vaxxers. There is nothing to be gained by taking it personally when student beliefs contradict well-established scientific knowledge. If we go out of our way to contradict a student by appealing to our authority as an expert, then we are shutting down the potential for learning. You can only succeed if you maintain open channels of communication, and if you provide information and evidence so they have the latitude to make up their own minds based on the evidence. Speaking in a derogatory or condescending manner about people because of their beliefs will undermine your own goals.

It is not recommended to set up a debate and "teach the controversy" so you have students compare the validity of arguments that are based on solid science and arguments that are not based on solid science. Since where you are coming from is clear to the students, it is not respectful to those who hold opposing views to set up a forensic competition against what you have predetermined is a losing side. You do not have to give credence to views that are not supported by sound science, but you don't have to highlight them just to make the point that they are misfounded.

The best way to teach evolution, global warming, plate tectonics, or vaccines is to simply teach them as you would teach any other topic. If you feel that it's an important topic that needs to be covered in more depth, then just cover it in more depth. There's no need to call out people who disagree, and there's no need to debunk nonscientific approaches to evaluating evidence. It is better to infuse your entire course with evidence-based decision-making, for all kinds of noncontroversial topics. Just like for every other topic, provide a lesson that involves inquiry and evidence to demonstrate that the key point you are working to make is the best tool you have. It's not reasonable to think that a one-hour lesson from any science professor can, on its own, change the mind of a person who is personally invested in thinking that the science is wrong. On the other hand, if you present the science from a sensible and respectful perspective, that helps open the door a little bit more for those who are positioned to walk through it.

What Is Academic Misconduct?

Every university has some form of academic integrity policy, which defines academic misconduct. I've seen a bunch, and they are pretty much the same: "don't cheat, don't let someone cheat from you, don't plagiarize or contribute to plagiarism, don't lie about attendance or assignments," and so on. You get the idea. There are also policies about what instructors are supposed to do when they encounter a violation of the academic integrity policy. This usually entails contacting an administrative office, usually the dean of students, but only after discussing the matter with the students. The consequences for the misconduct usually are left up to the individual faculty member. The institution will keep tabs on what happened, and at least in theory, if there are multiple incidents the university will take further steps to penalize the offenders. (In my experience, this rarely if ever happens, even when multiple incidents are on the record.)

Some campuses have an official honor code, spelling out explicit policies and consequences for honor code violations, which are handled by a campus honor council. (I once taught at a college that required students to handwrite a one-sentence honor code statement on all assignments, including quizzes, and faculty members needed to be out of the room whenever assessments were happening!) Honor codes don't seem to change the classroom dynamic that much, as the limited research that has been conducted suggests the rate of cheating is similar between schools with and without honor codes.

In chapter 2, I suggested that you put in your syllabus that academic misconduct will result in an F for the entire course. If students are informed about this policy when they start your course, then they clearly must be aware what will happen to them if they are detected cheating or intentionally plagiarize. I usually get two kinds of responses from people when I suggest an F in the entire course. Either it's "Of course, that's the only choice," or "Oh my gosh, that's so extreme!" I've worked in some places where the consequences have been severe, and another place where the consequence was relatively light (such as a score of 0 for the test the student was cheating on). While strong consequences are not a deterrent, light consequences essentially are an encouragement. Keep in mind that misconduct also impacts honest students, and not just by affecting the grade distribution. If misconduct is rife, then those doing their work by following the rules will be aware that others are cheating, and this is quite unfair to them.

Discovering academic misconduct in your class can quickly turn into a minefield if you don't navigate your way through very carefully. For nearly all of the parties involved, the reputational stakes are high. Students think their life will be ruined if they are busted, and especially if this is a high-tuition private school, parents and lawyers can get involved. Faculty members don't want people to think the department has an integrity problem, and it can involve additional meetings and bureaucracy for chairs and deans. To help you navigate these difficult waters, I'd like to introduce some ideas to put these situations in perspective.

At most institutions, there is a vast canyon separating the written policies about academic misconduct and how faculty members actually handle misconduct incidents when they occur. The written rules have little bearing on cultural norms. The number of reported incidents is far, far lower than the number of observed incidents. Does this mean that professors are just going rogue and breaking university policy all the time? Probably. If you go ahead and follow the letter of the law every time you detect academic misconduct, this might put you in conflict with more experienced instructors in the department who do otherwise. It also might be best practice and a wise decision. Before you take any action, it would be useful to consult with the person up the chain of command who you report to, to make sure that your actions aren't going to create bigger problems for yourself.

In earlier years, when I encountered academic misconduct, I used to be bothered by the poor judgment of the students. If students are driven to cheat, how can I be an effective instructor? What did I do wrong? I've gotten over that worry. It was cured with data. Cheating on exams and assignments is rampant in all sectors of higher education wherever you go. That's just a straight-up fact. Every class you're teaching, there's someone cheating in it. About three-quarters of all college students cheat during their undergraduate years, and this cuts across gender, income, academic ability, grades, and institutional prestige. Even though I rarely see it, I don't harbor any delusions. Some students are breaking the rules.

There are many students who sincerely don't view cheating and plagiarism as major ethical breaches. Whenever I tell people that academic misconduct is very commonplace—and share the research results showing that this is the case—I get a lot of resistance from my peers, who are reluctant to believe that their students could be breaking the rules so flagrantly. This isn't so extreme to the students, though many of them don't perceive this kind of behavior as unethical, because they feel the entire system is just an unfair and adversarial game rigged against them, and to some extent, they're right. Most programs and most courses are designed with a huge incentive to cheat. Keeping the Respect Principle in mind, I think it is important to consider an incentive system for academic misconduct from your students' perspective. As always, keep in mind that your mindset toward higher education as an undergraduate may have been atypical, considering you ended up teaching at the college level.

When you are keeping the perspective of your students in mind, this includes the students who are rightfully indignant if they see their professors are not taking misconduct by other students seriously. In an environment where cheating is overt and rampant, students who go by the rules will feel at a disadvantage and that you are not being fair to them. And they would be correct. An application of the Respect Principle means that you have to take misconduct seriously.

In light of Efficient Teaching, it is wise to take steps to minimize cheating, by not only reducing the opportunity but also reducing the benefit to be gained. Handling matters of academic misconduct perhaps might be the most massive time suck you will experience while teaching—even more than grading a tall stack of long essays. It may involve more work and worry than you might imagine. If you're dealing with anybody other than the students, there will be at least some paperwork, and maybe some follow-up emails and/or meetings. While dealing with the student, you'll need to take the time to choose your words very carefully, and it's common that the student will want to involve other parties if they claim they are not being treated fairly. If I do feel bothered when a student is cheating, I don't take this as a personal offense. It is most unfortunate because it takes away my time, which means less time is available for the course and all of the other students. Cheating harms the students who aren't cheating, not just by creating what might be an unfair advantage, but by misplacing the time of the instructor who should be focused more on teaching than on handling misconduct issues. While I'm not suggesting that you choose a route that means the least amount of work on your part, you might be aware of what you're getting into before it takes up a lot of your time.

The best way to deal with academic misconduct is to prevent it from happening in the first place. While it's not a bad idea to include a clear and unequivocal statement of the consequences, research shows that the severity of the consequences doesn't necessarily have a big impact on the frequency of misconduct. I'll describe what you can do to minimize the incentive and opportunities for cheating on exams and how to minimize plagiarism. But since both of these things may still happen anyway, I describe ways that you can handle these scenarios when they emerge.

Minimizing Cheating on Exams

The best way to deal with cheating is to remove the incentive. You can't do this entirely, unless somehow you get rid of grades, but there are some general approaches and very specific steps that you can take to prevent academic misconduct.

Keep in mind that when students make a choice to break the rules, this is presumably because they're focused on the outcome—making sure they don't get a grade that they find unacceptable. By the time this happens, the notion of learning has gone out the window and students are just focused on jumping through the hoops that you've prescribed. This might sound hokey and naïve, but if you manage to create a learning environment where students believe in the value of what they are learning and respect you as an instructor, then you might have less cheating. When students feel they are being treated fairly and respectfully, fewer will resort to misconduct.

Anxiety about grades might lead to cheating, and to some extent, we can help reduce this anxiety. When there is a massive amount of points on the line, anxiety grows, so if you can reduce the importance of any single exam, it's less likely that you'll see cheating. One way to do this is to drop the lowest score on every category of assessment, as discussed in chapter 2. For writing assignments, when we provide a very specific grading rubric, students will have greater confidence in how the assignment will be graded. Also, the more specific we are about the kind of information that will be on a test, the less anxious students will be on the day of the test. Of course, there are plenty of factors outside our control. Students angling for admission to competitive professional programs that use grades as a filter (for example, premed students) are definitely anxious about their grades, as are students who have a GPA threshold for scholarships.

You are probably familiar with the standard tips and tricks to reduce the temptation for cheating. You could use multiple versions of an exam, require students to deposit book bags in one spot in the room, and space students out as far as possible. Another tactic is to simply allow an open-book text and write the exam with this in mind, or to allow students to prepare and bring in a single sheet of paper with their notes on it.

Considering how common cheating truly is, this might lead you to consider the educational value of administering exams that don't allow students to access sources of information while taking the exam. The bottom line is that if something is *verboten* while you're administering an exam, at some point you'll have students working to circumvent this rule.

Handling In-Class Cheating

Ideally, you and your students will know exactly what will happen when you detect cheating, because the consequences are in your syllabus. If your syllabus is clear, then taking action is simply a matter of implementing your policy.

Once you have determined a student is cheating, I recommend that you inform the student in writing in a memo on departmental letterhead about your finding, being specific about the consequences and the evidence of cheating. Then, there is a process to notify the appropriate office (perhaps the associate dean, or the dean of students), and be sure to let your chair know as well. Ideally, at this moment, the matter has been put to bed.

It's possible a student will dispute your decision through official university channels. This process is easier if your syllabus already clearly addresses the consequences of cheating, but not having a syllabus policy for cheating doesn't mean you can't flunk the student for cheating. If the student wants to meet with you, it's important to make sure that a third party is present (another faculty member would be best). It also helps to let at least a day go by after you notify them of the consequences, because this will give them an opportunity to choose against a rash course of action that is against their own interest. In every university that I am aware of, if a student disputes a faculty decision, then the student will need to demonstrate how the faculty member is wrong. In other words, if it becomes a you-said-they-said situation, the default is that you are correct. With this power comes responsibility, and, of course, unless you are unequivocally certain that misconduct has happened, it's wrong to accuse a student of cheating.

There are a few ways to determine that a student is cheating. The first one is to simply observe a student accessing unapproved information during an exam. This should be really straightforward. But on the other hand, sometimes when you are grading exams, you might unexpectedly discover parallels between the responses of one student and another student. It's hard to rule out coincidence and study partners, but once in a while it might be clear as day, including who copied from whom. It would be wise, in this kind of situation, to run the exams by trusted colleagues who agree with your judgment. A third way you might discover cheating is when you are informed by a third party that it happened. Unless you have some evidence that is not hearsay, you should not take any action. However, this might be a cue to pay particular attention the next time you administer an exam.

Minimizing Plagiarism

Your idea of academic honesty is probably very different than what your students have in mind. I think everybody has been given a clear definition of plagiarism at multiple points in their academic career and been told how to avoid it. But if you've gotten this deep into the book, you are well aware of the notion that being told something doesn't mean that someone has truly learned something.

I think many students who are plagiarizing are genuinely not trying to game the system but sincerely are not familiar with how to draw from sources appropriately. Admittedly, it is inevitable that students who are intentionally plagiarizing will claim that they did not do it on purpose. If students are seeing their coursework as a game to be won, from an adversarial perspective, then plagiarism can be seen as merely one strategy available for playing the game. I think one way we can work to minimize plagiarism is to engage with students in good faith and avoid pedagogical approaches that support treating your course as a matter for gamesmanship. Students are aware that plagiarizing is an ethical decision, and one of the major rationalizations for this form of misconduct is that their professor has set the course up as a game of hoops to jump through. If you develop the writing assignment in a way that genuinely helps the student learn what you're trying to teach and is equitable, then they might feel less incentive to plagiarize.

How can you tell the difference between intentional and unintentional plagiarism? The bottom line is that you can't. There are always going to be people who are working to game the system and have no problem with breaking the rules, even if we develop a trusting environment for our students. Some are betting on the low risk of getting called out for plagiarism and counting on a slight penalty if they are caught. If a student says, "I had no idea this was wrong!" then you aren't going to be hooking them up to a polygraph. You don't want to put yourself into a situation where you have to decide which of your students are being honest and which of your students are lying to you about unintentional plagiarism (as is also the case for reports of dead grandparents, in chapter 2). There are many things you can do to minimize the incidence of plagiarism, based on how you design the assignment. If you simply use your Spidey Sense to decide who is telling the truth, you're putting yourself in a position where you don't respect your students. On the other hand, if you entirely ignore the prospect of plagiarism in your classroom, you are disrespecting the effort of all of the students who are working honestly.

It is possible to treat plagiarism as a violation of academic integrity in a way that is fair to all students and also to require honest academic work. This starts with putting in an earnest and substantial effort to make sure that your students understand plagiarism from the get-go. If you're having students turn in assignments that have the capacity for plagiarism, then you owe it to your students to teach them what plagiarism is. For some students, this might be entirely review material. But students who don't know how to not plagiarize are unlikely to volunteer this fact, and it's better to provide this lesson to all undergraduates yet again, even if it is redundant with prior lessons in other courses.

Because plagiarism is so widespread, I don't think it's possible to overdo education on what constitutes honest scholarly writing. We can't just show up in class with a handout and a short lecture and expect our students to learn how to not plagiarize. If you're not keen on spending a lot of time in class on this, one way to deal with this issue is to spend a few minutes discussing the importance of this issue and the consequences for academic misconduct, and then you can refer your students to take an online plagiarism tutorial. There are some very well-designed training modules that directly engage students in activities that show them how to not plagiarize and that demonstrate how certain approaches to writing result in plagiarism. You can require students to show evidence that they've completed the module, are capable of identifying and avoiding plagiarism, and understand the consequences of plagiarism in this course.

It's likely that you have a reference librarian who has been assigned to your department. Their job is, in part, to work with you and your students on working with the literature. If your course involves a substantial writing component, I think working collaboratively with a reference librarian from the outset of the course can be helpful, and from a more pragmatic standpoint, it's more efficient for the expert in referencing literature to work directly with your course on teaching this material.

Some instructors go to extraordinary lengths to document that their students have been taught the distinctions between plagiarism and honest scholarship, by asking students to sign a contract that specifies they will not take steps that will result in plagiarism. The notion of this contract is to make sure that students have actually done the online training, to make sure that they sign their name to a paper that they acknowledge what plagiarism is and that they understand the consequences if that's a choice they make in your course. While plenty of your students will not be convinced that plagiarism is unethical, you can make sure that they all know what it is, that it is against the rules, and that it will be treated as a major breach.

I don't recommend the contract approach to plagiarism. I acknowledge that it provides a close-to-airtight method for busting students who have violated your policy, but keep in mind that this also places you and your students in an adversarial posture against one another and heightens the atmosphere of gamesmanship. Having students sign a contract that they won't plagiarize is a tacit recognition that plagiarism is routine in your course and that if students violate your policy, you can whip out this contract to throw the book at them. I think it's more simple to just put a sentence or two in your syllabus that students are taught what plagiarism is, that they are expected to not plagiarize, and that this is regarded as a matter of academic misconduct. Providing actual lessons on how to not plagiarize is more important than having students sign a contract saying they won't do so.

The most important step you can take to deter plagiarism is to scaffold writing assignments so students are required to take the steps involving honest scholarship. In a major written assignment, you can require students to develop a short summary, provide an annotated bibliography, submit an early draft, and then turn in the final draft. This will not prevent students from plagiarizing, but it also gives them more of an opportunity to do honest work and takes away the need to do a large project from scratch to finish in a short period of time. Regardless, scaffolded writing assignments are just a good approach to teaching writing. This also might make plagiarism seem more obvious when it crops up in the final product, which might deter students from doing it. In the notes at the back of the book, I suggest two books that provide a more detailed approach about teaching well in a culture of cheating and how we can structure our institutions and our courses to minimize its impact on student learning.

Handling Plagiarism

Let's say you detected plagiarism, one way or another, and let's say, based on the information available to you, you are wholly confident that the student should bear the full responsibility for their actions. To bring the Respect Principle into play here: We respect our students by giving them the benefit of the doubt. So, if we decide that they plagiarized intentionally, that means there has to be enough evidence and context to overcome the doubt.

Now that you have made such a determination with fairness and a generous heart, how do you handle this? You'll need to sort out the consequences that you think are most appropriate. Unless you're well accustomed to dealing with this in your current department, it's a good idea to talk to an experienced instructor in the department. It's alluring to let the incident of plagiarism slide with minor consequences, because if you treat it seriously, the student might choose to make a big stink. However, I've seen too often how treating intentional plagiarism as a minor incident invites more plagiarism. Once you get a reputation for taking plagiarism seriously—and the reputation does get out there—you have a deterring effect. Keep in mind that if you treat plagiarism differently than everybody else in your department (either with less severity or with more severity), there may be a cost to you for not conforming to social norms.

When overt plagiarism is treated lightly, academic misconduct flourishes. Students who plagiarize successfully are more likely to plagiarize in the future. Let me give you an example. In a department where I worked, a professor assigned a major term paper about viruses. One student chose to write about Ebola viruses and produced a very thick end product, going on for many pages. The first page or two looked solid, but then it took a strange turn. The paper started talking about how Ebola viruses can result in adverse consequences in the home, and then it discussed the dangers of frequently drinking "Ebolaic beverages." After a quick search, this professor sorted out that the student took a journal article about alcoholism and merely did a search-and-replace to change "alcohol" to "Ebola," to fill in several pages of their term paper. The student even invented fake journal articles for the reference section! I still shake my head, thinking, "What was this student thinking?" But really, we all knew what the student was thinking. At the time, in this department, there had been a history of letting plagiarism slide by. In the proceedings that followed this case, the student reported that they had done "research" in a similar fashion in the past, and it had been acceptable to their other professors.

I can understand why some faculty gloss over intentional plagiarism, because the more you look for it, the more you see it, and dealing with it often creates misery and pushback. I also think that the more we play the cop and the judge, the less we teach those who are there to learn.

Let me first give you an example of what not to do. The first time I intercepted academic dishonesty as a new faculty member, I was teaching a well-established lab section for the first time. I saw that five students, who were writing papers on a major research project in the lab, were all plagiarizing together. It was overt and badly done, and they clearly knew it was against the rules. One student even came to my office hours with a paper written by a friend from last year's course (with remarks from the prior instructor still on it!) to pass off as their own and ask me for input. I confronted the students and, backed into a corner, they lied and denied it. The facts of the situation overtly contradicted their claims. You would think a student couldn't just claim that their paragraph matching a paragraph from somewhere else could be mere coincidence—but they did! I spent days—literally—doing all the legwork, paperwork, documentation, interviews, and other crap associated with making sure these students got their fair dose of justice. My department was very supportive, and my chair expected me to handle the case this way, but it mushroomed, because the students escalated the issue. It turns out that in my department, students weren't used to getting busted for plagiarism. Some kind of hearing was held. It was a big deal. Nobody wants to go through this. It turned into something resembling an inquisition, because it promptly became an adversarial situation.

I really screwed that situation up. I didn't even imagine how so many folks could so obviously lie about dishonesty, in the face of incontrovertible evidence. They'll claim it's just a coincidence. Even when this defies all logic. When students fight against your claim of plagiarism (even though evidence obviously says otherwise), this necessitates a process that will cause you more paperwork and trouble with administrative offices. They're making you do it the hard way. In the end, this strategy might work for them. It's amazing how consistent denial in the face of clear evidence can persuade people who want to see the problem go away.

When we directly accuse our students of plagiarism, they'll go defensive and deny it. This then puts them into a hole where they've already lied, and then they feel like they can't recant their lie, because then they'd be guilty of lying. So, they might just continue to make an absurd claim, rather than capitulate in an adversarial situation. For this reason, I recommend against directly confronting students in person with intentional plagiarism, because it makes the situation more adversarial than required.

Most university policies ask that you raise the matter of plagiarism with the student before you report it to the university, so that if there is any misunderstanding the student will have an opportunity to clear things up. Of course, I think it is important that a student have an opportunity to explain themselves before you report them for misconduct. It's better for everybody if the student has an opportunity to reflect on your finding and the evidence before responding to your claim, and you can do this by handling this matter in writing instead of in person. I suggest writing a memo in which you explain the evidence for plagiarism and inform them of the consequences. Then tell them that you are planning to report this to the university and you would like to give them a day to consider the circumstances before responding in person or in writing. If they'd like to meet with you, make sure that it's scheduled for a time when another member of the faculty can be present. I make a point to remind the student that I am required to follow this policy, that I want to represent their interests as best as possible, and that I do not plan to request further sanctions from the university. If they track you down and try to talk without arranging a meeting with a third party present, tell them that you aren't available to talk and that they should put down their response in writing.

When you inform students about this by email, you might just never hear back, and you can just levy the consequences and move on. It's also possible that the student will want to meet with you to proffer an explanation, involving excuses or mitigating factors, and of course, if you are wrong in a way that you didn't anticipate, this is when the student can set the record straight. I think it's important to treat all of our students with respect and kindness—including the ones who have violated academic misconduct rules.

I would like to point out that I will only levy sanctions for plagiarism under two circumstances. First, I must be sure that the student is fully aware of what plagiarism is and was legitimately trained how to write without plagiarizing. Second, I must be sure that the student should have been fully aware of the consequences of plagiarizing when they received the assignment. In other words, I treat plagiarism as seriously as cheating, but only because I have been perfectly clear that plagiarism is as bad as cheating on an exam. If you are unsure that your students are prepared to do honest scholarship, then it would be unfair to punish them for breaking the rules if they don't know how to not break the rules. A lot of college students emerge from high school with no real experience writing without plagiarizing. If we have high expectations of our students without giving them the resources they need to succeed, then we are just part of the problem.

8

Online Teaching

Online Teaching Is Still Teaching

So, you have to teach online, but aren't sure you're ready? The good news is that the fundamentals of teaching always matter. Both the Respect Principle and Efficient Teaching are just as relevant when teaching online. Healthy instructor-student relationships often organically emerge from daily interactions. When we share a physical classroom with our students, it can be easier to build rapport and humanize one another. However, when we don't share the same physical space, we need to make a deliberate effort to foster mutual respect with them.

Some popular notions of online learning are misleading. MOOCs (massive open online courses) are not typical. Neither is the emergency online teaching that happened in spring 2020 during the COVID-19 pandemic. In effective online teaching, instructors actively engage with all of the students enrolled in the course. Even though we are not in the same classroom as our students, online teaching can be just as personable as a face-to-face class. We just need to be intentional about it.

Online Teaching Is Different

Teaching is still teaching, but online teaching and face-to-face teaching build from different instructor skill sets. We should not be teaching online the same way we do in a classroom, because what works in one instructional mode may not work in the other. I like to think of this not as a bug, but as a feature. While we have extraordinary freedom when we are teaching in our classrooms, we are also limited there by boundaries of time and space. When a course is online, we are liberated from these boundaries and can engage in some teaching practices that are enhanced by the online environment. There are many ways for students to learn that are well suited to places beyond the classroom, including careful reading, problem-solving, creating media, and other projects. Part of building an effective online course is to avail yourself of modalities of learning beyond the traditional classroom lesson.

Another major upside of online teaching is that it generally follows the pace of each student. For example, if a student is watching a video, they can pause when they need to, and they can go back and watch a clip as often as they want. If students are having a discussion in an online forum, they have the time to be introspective about what they write, and to respond when they have fully formed their ideas. You can worry less about going too slowly for those who already understand a piece of content, or going too quickly for students who didn't catch it the first time.

Online teaching allows flexibility in how we package our lessons. It's hard to design a classroom lesson that lasts the precise period of time that we are expected to fill in a given day. Since we can't perfectly time all of our in-person lessons, we end up being less effective when rushing to stay on schedule or finishing earlier than expected. In online teaching, the lessons can be chunked into sizes that naturally fit the course material. For each week, you'll have a certain amount of content on the agenda, and it's okay for it to be a little longer sometimes and a little shorter other times.

It's easier to teach in a media-rich fashion in online courses, and it's easier to integrate hands-on digital work into your lessons. If you're teaching in a classroom, perhaps not every student has a laptop while you're teaching, and it's clunky at best to pause while students are asked to load a particular website or download a particular piece of software. However, in online teaching, you can have students do an activity, and then return to the lesson. While doing a think-pair-share isn't possible when students aren't taking the class at the same moment, you can design online courses so that students can think and share online before moving on to the next piece of the lesson.

Characteristics of Effective Online Teaching

We all have been students in high-quality college courses, taught by faculty who really knew what they were doing. However, many of us have never taken an online course, and even fewer have taken one that was well designed. Unless you're intimately familiar with the genre, I highly recommend taking the time to investigate existing models of good online courses. Your institution's Center for Teaching and Learning likely has some to share with you, and you can check out resources from the Association of College and University Educators (ACUE) and Quality Matters (QM), an organization designed to ensure the effectiveness of online teaching. There are many different ways to run an excellent online course, so you'll probably see a lot of variation. Of course, regardless of mode of delivery, all courses need solid curriculum, pedagogy, and student assessment. On top of that, there are several course design aspects that particularly matter for online courses.

Protecting student privacy is important, and is often a matter of legal compliance. You need to make sure that student grades and your remarks on student performance remain private. The LMS (Learning Management System, covered in chapter 4) software is built with these privacy needs in mind, but it's important to not insert new elements into your course that might breach student privacy. For example, students should not be required to display video with their own homes in the background, or asked to reveal private information about grades that they had received on prior assignments.

Effective online courses are built so that it's easy to navigate the LMS. In an online course, the kinds of details that are typically in the syllabus should be spelled out prominently and in adequate detail, so that students are aware how to earn full points for the course. Not all of the students will be acquainted with the operation of your course in the LMS, so a concise document that lays out how to access all aspects of the course is important. Even better, you could record a video with a screen capture of a narrated tour through the LMS. You can walk students through and demonstrate how to find all of the class materials, where to participate in discussions, how to submit assignments, how to see their scores from assignments, and how to do collaborative work.

In addition to reviewing examples of high-quality online courses, I suggest that you use an established rubric to ensure that your own course isn't missing important pieces. You can find them through ACUE or QM, and many universities develop their own as well. Many of their criteria include elements that I've mentioned in this section. These standards will help you develop the elements of your online course, some of which might not have occurred to you.

Asynchronous or Synchronous?

Synchronous instruction is when everybody in the class logs in at the same time for regular class sessions. Asynchronous instruction is self-paced, within limits, with lectures and other materials provided online for students to access at their convenience. Even though synchronous online courses provide a vague verisimilitude of a classic classroom experience, asynchronous instruction is very common in the online teaching universe.

If you try to teach a synchronous lesson online just like you would be teaching in person, you and your students are likely to find the experience to be subpar. However, if you approach a synchronous lesson in a manner that is well geared toward online learning, then there is no reason to expect it to be less effective than an in-person class. Online teaching isn't better or worse, it's just different. If we focus our educational efforts on an attempt to recreate the classroom experience in a virtual environment, we are limiting our options and not availing ourselves of the effective asynchronous pedagogical approaches that can be deployed to great effect in online courses. The factors that can help you choose the synchronous and asynchronous elements of your course often involve logistics, suitability of content, and the needs of the population of students that you are teaching.

Teaching is an art, and synchronous and asynchronous teaching are different art forms. If you have developed a talent for teaching synchronous face-to-face courses, then shifting to asynchronous online courses might be like asking a sculptor to produce an oil painting. Both the principles of teaching and the principles of art transcend medium. Nonetheless, to produce a piece of art or teach a good class, you'll need to be able to use the tools particular to the medium and understand the strengths and limitations of different approaches.

Let's consider the positives of synchronous class sessions. Building a community of learners is a part of effective teaching. Having everybody together in a virtual "classroom" provides opportunities to build such a community. Synchronous lessons allow you to split students into small groups to work on problems together, which is facilitated by software that is probably built into your LMS. Synchronous lessons help you and your students compartmentalize your time. If you are teaching a population of students who attended college straight from high school and are principally experienced with in-person learning, they are less likely to have experience managing their time in an asynchronous course, and synchronous courses will help them regularly focus on their learning. Synchronous teaching elements also allow students to be more present with their full identities, which helps build a more inclusive community for students of color and disabled students. Synchronous lessons are not an all-or-nothing endeavor. It's normal to blend synchronous sessions with asynchronous elements. When you are starting your course, you can learn from your students whether they find synchronous class sessions to be helpful, and you can always offer informal synchronous meetings as an optional course element. When you hold synchronous lessons, it's helpful to record them for secure viewing on the LMS by students who were not available at the original time or would like to review the lessons later.

To understand the benefits of asynchronous instruction, it helps to consider the usual reasons students take a course online: because it's more convenient, or because they don't have a choice. While some students simply prefer an online class, many are not available to be in a classroom, and online courses offer a practical alternative. The same factors that might prevent your students from taking classes in person—such as work schedule, family care responsibilities, or limited access to a reliable internet connection—may also prevent them from participating in regular synchronous sessions.

One benefit of asynchronous courses is that their flexibility allows students to do schoolwork when they are at their best. Face-to-face courses are conducted at the time and place chosen by the university. Some classes are in the afternoon doldrums, and early start times can bring in underslept students and harm student attendance and engagement. On any given day, some students will have academic or personal challenges competing with their readiness to learn. In asynchronous courses, students can choose times when they are in a physical and mental condition to perform hard cognitive work. If you can give students a choice about when to log on, they're likely to do so at times when they are primed to succeed.

Keeping Students on Track

One drawback to the flexibility of online learning is that it can be harder for students to stay current with assignments. We can intentionally design our courses to help students stay on track, and give them leeway if they get a bit behind. In online classes, it's common for a student's final grade to be based on a combination of many small quizzes and assignments instead of huge midterm and final exams, which can be challenging and impractical in this setting. Students can take the smaller quizzes every week or every two weeks. This allows you to assess how they are keeping up with the course material, and a bad score on a smaller quiz hurts their grade less than one on a major exam. Breaking large assignments down into smaller intermittent pieces can also help, but only if you keep tabs on who is turning them in and reach out to students who are behind.

In addition to helping students who are struggling, we can help our students by creating courses that are accessible and engaging, so that they are less likely to fall behind. As discussed in chapter 4, students learn better when engaged in active learning and working with other students to solve problems and think critically. This is true even if students are interacting with classmates electronically. If students are treating your course as a solitary endeavor, their isolation makes it harder to learn. On the other hand, group learning is well suited to online courses, and you can help students be accountable to others by creating opportunities for them to engage with fellow students while completing their assignments. Students are more likely to keep up with their work when it is a genuine community experience.

Building a Community

Colleges are great places to learn because they are immersive communities of learning. Notwithstanding the achievements of autodidacts such as Frederick Douglass and Bill Gates, there is copious evidence that colleges teach more effectively when students work together. To preserve this strength, we should teach so that our online class is more than an aggregation of students who are learning in parallel. How do you create a community in an online environment? How do you do this if the class isn't meeting together frequently? Experienced instructors of online courses have a wide range of practices.

How we set the tone at the beginning of the course has a huge influence on the rest of the semester. Students should have an opportunity to introduce themselves to you and the rest of the class. While you could merely ask them to write a little bit in a classroom chat board, it's more impactful to ask all students to share an optional 1–2-minute video in which they introduce themselves. You could model how this is done by uploading your own video from the camera on your computer, showing that it doesn't have to be a work of art. You can ask students what names they prefer to go by, any interests relevant to the course, anything about their background they wish to share, and perhaps a fun question of your choice. This might seem a bit awkward as an "icebreaker" activity, but it really will help everybody connect names to actual people.

When professors show more of their human side by discussing their interests and hobbies, it opens up space for development of an engaged learning community. It's okay to let interactions move into non-academic areas, as long as they're not disrespectful or polarizing. This practice allows everybody to show a bit of themselves, which is part of community-building. You can open up a discussion board in the LMS specifically for students to discuss things unrelated to the course. You also can set up social media accounts for the course, where you share how the course material intersects with life in the outside world. This can result in comments and conversations among students, creating a dialogue beyond the borders of the digital "classroom."

Even if you're running an asynchronous course, it can be useful early in the semester to schedule some synchronous video chats. Make sure these meetings have an educational purpose that is fulfilled by bringing people together. If you have students bringing in their work from problem sets, or readings to discuss, you can assign them to virtual "breakout rooms" to work together.

Students will engage more with one another if work is done in smaller groups, rather than in a class-wide discussion. You don't have to create a dreadnought-scale "group project" to have students work together. Instead, you could have students comment on one another's drafts, or work on individual small assignments, or have a discussion about a piece of writing or a case study.

You also can encourage students to work together by having them create a single comprehensive set of crowdsourced notes. You can set up an online document that all of the students can edit and refine, just like a wiki. Clearly, some students will be contributing more than others, but it does help students feel a sense of community when they see other people who are part of the same experience building an understanding together.

It's common in online courses for instructors to allocate a bigger fraction of the grade to "participation." In face-to-face classes, you can run an active learning environment where all students are engaging with one another frequently, so participation points won't necessarily support student learning. However, in an online course, when students don't interact, they are essentially invisible to their classmates. While I've argued against "participation" points for face-to-face classes in chapter 2, I think they can be useful for online courses. If points are what it takes to bring a student to engage with the course community, this builds a more inclusive learning environment. It's okay to post questions to forums in the LMS and require students to post replies and engage with other students. Ideally, you will be able to design a curriculum that organically brings students to work together.

Equitable Course Design

The Respect Principle requires that we work toward educational equity. You are likely aware of the bad news that student ethnicity and socioeconomic status are associated with gaps in course grades and graduation rates. Here's some worse news: these gaps are often amplified in online teaching. Now that we are aware that this is a potential pitfall, we can make equity a priority and develop courses designed to promote success for all of our students.

To make choices designed to eliminate inequities, we need to understand what's driving disparities. Students working from home might not have access to all institutional resources that are designed to address these gaps. Moreover, working from home is often a greater challenge for first-generation college students and those with low socioeconomic status, because they are less likely to have their own quiet place to focus on schoolwork at home. Another factor is that student diversity is less tangible in the online environment. While our goal is to work with the whole student, barriers to interactions in online courses can prevent us from understanding where our students are coming from, so it's more difficult to change our actions to meet their needs. The way to address this problem is to use a lens of equity throughout all course design, acting as if all students are facing barriers unknown to us, so that our course can properly eliminate them even if we don't know which particular ones a given student is facing.

It helps to be aware of assumptions that we might have of our students and disabuse ourselves of them. Students might be working more hours than we are expecting, and they might have responsibilities to care for family members. They might not have a laptop that can handle computationally intensive software, or one with a good microphone or camera. They might not have a reliable internet connection, or transportation to a place that does. Because many students will be looking at courses on their mobile phones, it's a good idea to make sure that your course is accessible on such devices. We can't assume that students will reach out for support when they're struggling in the class. It's possible to design our courses to ameliorate these challenges, to make sure that all students have access to the full educational experience and an equivalent opportunity to learn and achieve. Equity and access should be a prerequisite of course design, rather than an afterthought.

In earlier chapters, I've argued that high-stakes testing and massive assignments worth a large fraction of points focus student attention away from learning. When a lot of points get decided in one fell swoop, students don't have many opportunities to recover from setbacks. This is particularly the case in online courses, and even more so for students from underrepresented groups, who may be experiencing stereotype threat. If your focus is on equitable learning, then your online course needs to have assignments designed so that students experiencing personal challenges still have an equitable opportunity.

As students from marginalized backgrounds are less likely to seek our help, we can be proactive to support student learning. If any student's work, or lack thereof, indicates that they are having some challenges with the course, it's on us to reach out to them. Some of the support that our students need might come from us, but they also have access to resources on our campuses that can help them, which they might not be aware of. Instead of suggesting a student contact the tutoring center, for example, you could take a more personal approach and connect them with the person in the tutoring center who is prepared to offer support.

Another element of equitable teaching is a clear roadmap for success. In addition to informing all students what they should be doing through regular reminders, you can also provide guidance for effective studying, including what kind of notes you expect them to make associated with each lesson. Be as transparent as possible about all of the elements that feed into the course grade, and clearly articulate these elements to the entries in the gradebook in the LMS. Students are often concerned about how their grades on individual assessments will affect their final grade. If you build the gradebook so that all of these elements are overt from the start, and maintain the gradebook well, this helps students stay on track. Because the LMS is often the main way that you interact with students, staying up to date with grading is a way to show students that you're focused on their learning.

When we build community, we create connections that buffer against inequities. When we help all members of the course to bring their whole identities into the community, we enable their success. We need to let all of our students know that they belong, and that all of them are capable of meeting the high expectations that we've set forth. It's okay to be unsubtle about communicating this with students. You can just come out and say, "You belong here. I value your work."

Managing Communications

I've heard a lot of academics discuss their email inboxes as if they are out-of-control firehoses. If you've needed an incentive to wrest control of your inbox, then online teaching might give you one. While you can and should conduct a lot of the business for your course through the LMS, make no mistake: you will be receiving and sending plenty of emails. While our academic culture tends to regard email as busywork and a thankless chore, I invite you to reframe email as a positive feature of academic life. Email isn't a single task; it's a means to communicate with other people. Email is how we plan for the future with others, schedule phone calls, get necessary information, and often how we support students.

The first step in managing course-related email is to minimize the number of unnecessary emails. If we can anticipate the information needs and concerns of students and proactively share this information, this serves students well and is a more efficient use of our time. If you've done a great job building a community, then active discussions in the LMS can be a way for you to address student needs in advance, and there, you can reach many students at once.

The second step in managing your course-related email is to start with a plan to communicate well with students in the LMS. I think it's important for our students to know that it's okay for them to contact us, and to be welcoming. It's common for students to not provide enough information or context for you to be able to address their question. It's not a bad idea to post an example of an email that provides all of the information that you'd like in it: full name, the course and section, the name or number of the assignment they're asking about, and a specific question that you can answer. If they have a more vague issue to discuss, perhaps email might not be the best venue, and you could advise them to contact you during office hours or by appointment if you prefer. Questions about course content are often well suited to a forum in the LMS, so that other students can also have the benefit of seeing and learning from the exchange. It's also a good idea to instruct students to use their university email address and not a third-party email account. Those email exchanges could get lost in a spam box.

Because online teaching involves a lot of emails, you will probably want to compartmentalize course-related correspondence. If you're using your inbox as a to-do list, stop it. Instead, once you see an item in your inbox that you don't immediately deal with, you can remove it from the inbox and place it in a to-do folder, or flag it appropriately.

You can manage email traffic by automatically labeling messages from student email addresses. When the time arrives that you have budgeted for reckoning with student emails, you can go through all of the emails from students. If you want to get a bit fancier, then your email filters could specify the email addresses of the students in the roster for each course.

You will notice that many of your emails will fall into particular categories, and while you are customizing your replies for every student, many emails end up sounding similar because they are about similar concerns. It's efficient to create email templates for yourself, to copy and paste to save you the trouble of reinventing boilerplate responses. While you might feel some kind of guilt for not wholly composing every email from scratch, there is value in responding to similar questions in a similar fashion.

For some kinds of business, a conversation is better than email. If you think your email reply would take a lot of explaining on your part, then you might respond to the student by prompting a phone conversation. Alternatively, if you think your answer would be more effective or more efficient by talking about it, you could dictate a response into an audio file and then send that recording to the student by email. (This is also a great way to provide students with feedback on writing.) If your email is about a specific concern that you think should have a paper trail, then email is best. But if a student wants your feedback or you want to spend some time discussing a particular question or problem, then recording a three-minute piece of audio feedback might be better than trying to type out the same thing over fifteen minutes. Sometimes friendly nuance comes with your own voice, and sending personalized messages in video or audio format helps build a better connection.

It's helpful to put course-related email responses on your daily schedule. Once per day is just fine. While most students would prefer that we respond to emails even faster, the general expectation is a response within 24 hours, not counting weekends and holidays. Whatever your planned response time is, you should communicate this at the start of the course, and if you are going to be unavailable for some period of time, it's wise to inform the course that you'll be out of touch.

Running a Synchronous Video Session

Running a synchronous video lesson (in Zoom or another conferencing platform) requires skills that are not entirely intuitive on the part of the students as well as the instructor. Our species evolved while interacting in person, so interacting in a video environment is inherently awkward until all parties adapt. Here are some suggestions to help run a video session smoothly.

It will take time to establish guidelines for using the interface. It might seem a waste of time to spell out what might be routine and obvious, but even if one or two people break from the norms, it can set a lesson off track, so going over the ground rules for engagement is important. You'll want to make sure that everybody keeps their microphone muted unless they're speaking. When students would like to speak, will you need them to click the "raise hand" button and wait to be called on? How do you want to use the chat box in the video session? Under which circumstances would you suggest that students have their cameras on? These are matters to address at the start of every video session until a full rhythm is established.

Because behavioral norms in many online spaces are not as respectful as you would expect in your own virtual classroom, it will take effort to establish proper norms that promote learning and create space for all students to be comfortable. You should define your expectations about how you expect sessions to run. Students might be anxious that they aren't doing this right, so reassure them that we're all going to have awkward moments, and that you don't mind if their cat ends up on screen or if they get interrupted by someone they're sharing space with.

Office Hours

It's common for online courses to have busier office hours than traditional courses. This makes sense, because when you are meeting your students face-to-face on a regular basis, students will have the chance to ask you questions as the opportunity arises.

What do office hours look like when you're not meeting in an actual office? That depends on how you run them. It's possible that your institution has specific standards or expectations for how you are supposed to be available during official office hours. Or, it might be left to your discretion based on the needs of your students and your course. I suggest that you take a poll of your course at the start of the semester to find the times that work best.

Your office hours can take place over different kinds of media, including drop-in video chat, text chat in the LMS, telephone, social media, and even email. The idea behind office hours is that you have a back-and-forth conversation with students, in which they don't have to experience the time lags that come with typical email exchanges. So, "email office hours" would be a period of time in which you respond to all emails as quickly as possible, in the form of a conversation.

Having office hours on a drop-in basis may or may not bring in traffic. If students are aware that they can schedule a specific time slot to speak to you, this might lower the barriers and increase the chance that they'll contact you. Your LMS might have a handy feature to schedule appointments, but if not, there are third-party scheduling apps (for example, Calendly) that you can link within the LMS. If you have one good conversation with a student, this can increase the chance that they might contact you more in the future during office hours. If your class isn't too huge, you could even require all students to sign up for individual meetings with you, and associate some points with this. However, if you choose this route, you should state this in the syllabus at the outset of the course.

In addition to scheduled office hours, I suggest holding a discussion in the LMS that you can use as "asynchronous office hours," in which students can ask you any kind of question about the course, and you can then respond to these questions in a manner that all students in the class will be able to see. You can just call this a question-and-answer forum. While some issues for office hours should be private, it's efficient to answer questions that might be of general interest to the whole class, whether the questions are about content or course logistics.

Testing

In chapter 6, I recommended alternatives to assigning grades based heavily on the results of high-stakes exams. While these approaches might be a little unconventional for a college-level science class, they are entirely common in online courses.

Many online instructors are concerned about treating students fairly while administering exams, and this means taking steps to make sure that all students are following standards of academic conduct. Attempting to administer a typical science exam from a face-to-face course just doesn't work online. Without any proctoring, you can expect that some fraction of students will be cheating on the exam by accessing unauthorized materials. (To be clear, this is also true in traditional courses, though the fraction is likely to be higher in online courses.) This is unfair to the students in your course who are doing honest work. Your students deserve an environment that isn't designed to incentivize cheating.

For starters, take the time to explicitly state that you expect honest academic work, as this reinforces honest behavior. Some approaches to minimize cheating during online testing are better than others. One tack is to use educational technology tools. For example, you can use specialized browsers within the LMS that lock down access to anything other than your exam. You can use automated video monitoring of students via webcams, and your university might even contract out a service that conducts remote webcam monitoring. There are ways of administering exam questions with strict time limits that don't give students enough time to consult external sources. You can also randomize the order of questions, which might hinder cooperation among students. You can use a large bank of test questions, provided by the publisher of a textbook or of your own creation, and randomly select a subset each time. Some textbooks come with comprehensive student testing packages based on the material in each chapter, with testing services designed to minimize cheating.

If you've read through the book up to this point, then you can probably guess that I recommend against designing your class to increase the surveillance of students in online courses. This creates an adversarial environment that harms the educational environment and might even increase the motivation to cheat. While technological solutions designed to foil cheating might be tempting to you as the instructor, this approach can work against student learning. If you principally see your job as a prosecutor or a judge, then educational technology during high-stakes exams might help you do your job. On the other hand, if you see your role as an academic coach who is promoting student learning, then I suggest alternative approaches.

If you feel you need to administer major exams, you still don't need technological hypervigilance. You can design your exams to be open-book tests, which in the context of online classes is not substantially different than assigning students a take-home exam. This approach might be best for upper-division classes in which students are doing higher-level work involving applying ideas to new contexts and evaluating concepts.

You can simply not have quizzes or exams, and the grade can be made up of other types of assignments that can be used to earn points (for example: problem sets, video presentations, reports and research papers, class notes, journals). As already discussed, having an assessment every week or every two weeks is a way to identify students who are falling behind, so that you can intervene before it becomes a real problem for them. If you increase the collective weight of these quizzes so that they take the place of exams, you're sparing yourself the ordeal of administering major exams.

Laboratories

In many science fields, "online laboratory" sounds like an oxymoron. One purpose of labs is to provide hands-on experiences, with exposure to instrumentation and specialized resources. For this reason, a lot of online science courses will have the lecture portion in an online format, but then will bring students onto campus for the laboratory part of the curriculum. I have to admit, as a biologist, I'm not much of a fan of online laboratories, because some pieces of our undergraduate curriculum can't readily be replaced in the virtual environment. Online labs can't really teach you how to run a polymerase chain reaction, build an herbarium collection, or prepare slide mounts of bacteria. That said, many aspects of laboratories are quite amenable to the online arena. Some traditional learning objectives for lab sections might be difficult to achieve, but others can be well suited to online teaching.

If the purpose of your laboratory is for students to learn how to think like scientists, then you can do this in an online lab. We are used to students in laboratories performing the scientific method: developing hypotheses, designing experimental methods, analyzing and interpreting results, and presenting their findings. All of these things can be done online! Your students can still experience the full arc of the scientific method. Students can design an experiment, and they can collect or be provided with data they can analyze and interpret. You can assign students to work in groups to do everything that lab groups would normally do, aside from being in the laboratory to collect data for their experiment.

If you're slated to teach an online lab, I will guess that this situation arose in one of two circumstances, which are polar opposites of each other. You might be in this position as the result of long-term planning and careful attention to curricular design. On the other hand, it could be out of slapdash necessity (such as in the COVID-19 pandemic in 2020). If you're teaching an online lab with ample planning, support, and resources, then I imagine you're well prepared beyond the scope of this book. On the other hand, if you're developing and teaching an online lab in a hurried pinch, here are some approaches to provide some fuel for thought.

There are an increasing number of virtual labs developed by academic organizations and for-profit vendors. This trend started with virtual frog dissections decades ago, but now every field has offerings, and some of them are quite excellent. For most disciplines, academic societies have websites that will point you toward virtual lab resources. In addition, there are many online simulations in which users manipulate and acquire data to be used to test hypotheses that they developed independently. These types of simulations can work well in a broad range of topics, such as ecology, introductory physics, physiology, and engineering.

Online courses are not well suited to teaching hands-on laboratory skills. That said, there are many ways to collect original data without having your students in a university laboratory. Just because you're teaching an online course doesn't necessarily mean that everything has to happen on a computer. There are vendors that have developed laboratory kits for all kinds of disciplines for the purpose of online laboratories, in which students perform their experiments at home. These kits come with lesson plans, so it's not just the materials that come out of the box—the curriculum does, too. Without going to such lengths, it might be possible to find or develop a project or an activity in which students can collect meaningful data without having to purchase an expensive kit. If you've switched to online laboratory teaching because of a crisis, I think collectively surviving the crisis takes priority over preserving every aspect of a traditional laboratory education. However, if you've been asked to build an online laboratory course, I think it's possible to be creative and resourceful, and to provide students with an opportunity to do actual science, even if it's not happening inside a university laboratory.

Afterword

The End and the Beginning

I hope the end of this book might serve as a jumping-off point to learn about the science and scholarship of teaching and learning. Education scholars have as much jargon, competing theories, and strong opinions as we do in our own disciplines. When you first jump in, the water might feel cold, but you get used to it. It's always more fun with company. If you gather some colleagues for a group to discuss science teaching on a regular basis, that will increase your learning.

I asked some colleagues, "If you have a single piece of advice for a person teaching their own course for the first time, what would it be?" They said:

> Plan ahead as thoroughly as you can, but be acutely aware that no plan survives contact with reality. —**Susan Letcher**

> Be confident and excited. You are the expert in the room and the students came to learn from you. I think it makes the students feel more comfortable and creates a better learning environment. —**Sarah Lawson**

Think about what the students will be able to do by the end of the course that will show that they have learned something important. —**Eldridge Adams**

Transmit your enthusiasm to the students. —**Martin Heil**

Trust yourself. When you're a new (or new to that institution) instructor, students know it and they'll try to test you and get you to change policies. You'll see other instructors doing things differently than you and think they must be right. You are the instructor in your course; you know what works for you. —**Shannon DeVaney**

Be yourself. Students will always know if you're faking. —**Helen McCreery**

If something isn't working, change it. —**Jessie Williamson**

Acknowledgments

I am grateful for mentors who have taught, inspired, and believed in me. Beth Braker embodies the teacher-scholar who puts her students first. Mike Breed has been the model of a scholar, parent, spouse, teacher, and mentor. Kamal Hamdan shows how much of a difference a single person can make, by focusing on what really matters. John Thomlinson has been an irrepressible spirit and unflagging advocate. Thank you to the readers of Small Pond Science, the ongoing project that gave rise to this book. Stephen Heard, Jerry Moore, and Rob Dunn provided input from the beginning that helped shape this project. This book exists only due to the support of my spouse and best friend, Amelia Chapman. I have been inspired by working on this book alongside my son Bruce McGlynn, a devourer of knowledge.

I've had the fortune of being able to learn about teaching from some of the talented teachers in my midst, including Mark Osadjan, John Commito, Nancy Freeman, HK Choi, Helen Chun, Kathryn Theiss, Karin Kram, Fang Wang, Sonal Singhal, and plenty of others, who have been exemplars of effectiveness. Thanks to Ed Ayala, Alice Boyle, Prosanta Chakrabarty, Shannon DeVaney, Leonard Finkelman, Nancy Freeman, Nicole Gerardo, Neil Gilbert, Martin Heil, Sarah Hörst, Morgan Jackson, Sarah Lawson, Susan Letcher, Helen McCreery, Jeramia Ory, Santiago Salinas, Peter Shanahan, Dan Sheehan, Matthew Venesky, and Jessie Williamson, who allowed me to share their thoughts. I would like to share my gratitude to the late Eldridge Adams for his remarks as well. He was a role model as a scientist and as a reflective, talented, and student-centered teacher, and he is missed. Thanks to Amelia Chapman, Rose Ferreira, Sarah Rigley MacDonald, and Emily Magruder for remarks on the manuscript. CSU Dominguez Hills provided funding for the sabbatical that led to this book, and thanks also to Brian Brown, Lisa Gonzalez, and the crew at the Natural History Museum who hosted my sabbatical. Thank you to Gaby Gomez for everything on a day-to-day basis. Christie Henry helped develop this project, before it landed in the capable hands of Mary Laur. I have learned a lot about teaching from the secondary teachers in the cohort of the CSU Dominguez Hills Noyce Master Teacher Fellows in the Los Angeles Unified School District. It's been several years now, but I won't forget the lessons I've learned from my son's teachers at Longfellow Elementary in the Pasadena Unified School District, who have been models of effective and compassionate teaching: Tyara Brooks, Carolyn Brown, Mel Chidester, Tamyke Edwards, and Johna Steinstra.

Notes

Chapter 1

As teachers, we are more effective when we build an environment of mutual respect, where we trust our students and are transparent about why we are teaching how we teach. This essay related to this topic is about teaching English at the college level, but I think it translates to science quite well: Jane Tompkins, "Pedagogy of the Distressed," *College English* 52, no. 6 (1990): 653–60.

Much of the scholarship about how students learn better in an educational environment where they are respected is connected to improving learning conditions for students marginalized by their identity, such as in Jim Cummins, "Empowering Minority Students: A Framework for Intervention," *Harvard Educational Review* 56, no. 1 (1986): 18–37. I think the principle that students who learn best in an environment of mutual respect may apply generally.

One particular aspect of respecting students is recognizing that they have different backgrounds and different experiences than ourselves. The case is made here: Amanda J. Zellmer and Aleksandra Sherman, "Culturally Inclusive STEM Education," *Science* 358 (2017): 312–13.

The high rate of food insecurity and homelessness experienced by students in the California State University system was documented in this ongoing internal study by the CSU Basic Needs Initiative: www2.calstate.edu/impact-of-the-csu/student-success/basic-needs-initiative/Pages/Research.aspx. The extraordinary financial stress of undergraduates is not endemic to the CSU but is found in all kinds of institutions, often right under our noses. This is well documented here: Sara Goldrick-Rab, *Paying the Price: College Costs, Financial Aid, and the Betrayal of the American Dream* (University of Chicago Press, 2016).

Bloom's Taxonomy has evolved substantially since it was proposed in the 1950s. One summary that describes the theory accessibly and in some detail is L. W. Anderson and D. R. Krathwohl, *A Taxonomy for Learning, Teaching, and Assessing: A Revision of Bloom's Taxonomy of Educational Objectives* (Allyn and Bacon, 2001).

Chapter 2

Student cheating on exams and in plagiarized written assignments is particularly common. Study after study reports that the vast majority of students have reported cheating, and the number who cheat regularly may be well over 50%. Two highly cited reviews in this field are: Bernard E. Whitley, "Factors Associated with Cheating among College Students: A Review," *Research in Higher Education* 39, no. 3 (1998): 235–74; and Donald L. McCabe, Linda Klebe Treviño, and Kenneth D. Butterfield, "Cheating in Academic Institutions: A Decade of Research," *Ethics & Behavior* 11, no. 3 (2001): 219–32. While both of those papers came out a couple decades ago, this recent paper summarizes more recent evidence for a high frequency of cheating and plagiarism: Sebastian Sattler, Constantin Wiegel, and Floris van Veen, "The Use Frequency of 10 Different Methods for Preventing and Detecting Academic Dishonesty and the Factors Influencing Their Use," *Studies in Higher Education* 42 (2017): 1126–44.

While specifications grading and standards-based grading are not the norm in higher education, these approaches may be particularly effective at supporting student success. These approaches fall under the broader category of criteria-based grading. The distinctions among different types of criteria-based grading are explored here: D. Royce Sadler, "Interpretations of Criteria-Based Assessment and Grading in Higher Education," *Assessment & Evaluation in Higher Education* 30, no. 2 (2005): 175–94. A book that makes the argument for specifications grading, and provides practical examples about how to implement the scheme, is Linda Nilson, *Specifications Grading: Restoring Rigor, Motivating Students, and Saving Faculty Time* (Stylus, 2015). Another book that serves as a how-to manual for implementing a standards-based grading scheme is Robert J. Marzano, *Formative Assessment & Standards-Based Grading* (Solution Tree Press, 2011).

Chapter 3

An authoritative and well-read guide to curriculum design, though not specific to the sciences, is L. Dee Fink, *Creating Significant Learning Experiences: An Inte-*

grated Approach to Designing College Courses (John Wiley & Sons, 2013). The principle of backward design was advanced by the advocates of the "Understanding by Design" educational planning approach, often referred to by those in the know as UbD. For more details, see: Grant P. Wiggins and Jay McTighe, *Understanding by Design*, 2nd ed. (ASCD, 2005). It is not hard to find copies of rubrics and worksheets online to help use backward planning for individual lessons. I've found them to be very helpful when developing new lessons from scratch.

Does covering material in depth rather than breadth increase learning? One study answered this question by evaluating how the depth and breadth of prior coursework had prepared students to succeed in future coursework: Marc S. Schwartz, Philip M. Sadler, Gerhard Sonnert, and Robert H. Tai, "Depth versus Breadth: How Content Coverage in High School Science Courses Relates to Later Success in College Science Coursework," *Science Education* 93, no. 5 (2009): 798–826. In this study, the authors randomly selected dozens of universities and evaluated the relative depth and breadth of the high school coursework of currently enrolled students in introductory science courses. They concluded that increased breadth did not improve student performance, and in one field (biology), greater breadth resulted in lower performance. They also found that an emphasis on depth in one particular topic at the high school level resulted in greater outcomes overall in introductory college courses.

I suggested that I learned Newton's second law more successfully because I was not required to memorize it but instead was allowed to write it down on a permissible "cheat sheet" to take into a physics exam. Some studies have found that students prefer to take open-book exams or those with cheat sheets, such as: Uri Zoller and David Ben-Chaim, "Gender Differences in Examination-Type Preferences, Test Anxiety, and Academic Achievements in College Science Education—A Case Study," *Science Education* 74, no. 6 (1990): 597–608. The impact of this form of testing might not be on directly enhancing learning, but by reducing student anxiety, which indirectly may result in improved outcomes.

Many aspects of curriculum design are associated by discipline-specific standards. For example, in my own field, many curricular decisions are tracking the priorities and perspectives in the "Vision and Change" report: Carol A. Brewer and Diane Smith, *Vision and Change in Undergraduate Biology Education: A Call to Action* (Washington, DC: American Association for the Advancement of Science, 2011).

As you might imagine, the amount of research involving biases in student evaluations of teaching is extraordinary. The cumulative evidence indicates that there are biases based on gender, ethnicity, and age, and that these biases are intersectional. One recent well-designed study documents the effect size of gender bias and also provides an introductory review of documented biases in evaluations: David A. M. Peterson, Lori A. Biederman, David Andersen, Tessa M. Ditonto, and Kevin Roe, "Mitigating Gender Bias in Student Evaluations of Teaching," *PLoS ONE* 14, no. 5 (2019): e0216241.

Chapter 4

Even though effective classroom management is important at all levels of education, a lot of research and discussion targets K–12 teachers. One well-cited book that approaches the fundamentals of classroom management for students of all age levels is Robert J. Marzano, Jana S. Marzano, and Debra Pickering, *Classroom Management That Works: Research-Based Strategies for Every Teacher* (ASCD, 2003). The more interactive work you do in class, the more important it is to be able to rein in discussions, and this becomes more challenging as classrooms get bigger. Approaches to active learning in larger lectures that are amenable to effective classroom management are discussed here: Diane Ebert-May, Carol Brewer, and Sylvester Allred, "Innovation in Large Lectures: Teaching for Active Learning," *Bioscience* 47, no. 9 (1997): 601–7.

The concept of Scientific Teaching is described in this book: Jo Handelsman, Sarah Miller, and Christine Pfund, *Scientific Teaching* (Macmillan, 2007), which was preceded by a brief and highly influential article: Jo Handelsman, Diane Ebert-May, Robert Beichner, Peter Bruns, Amy Chang, Robert DeHaan, Jim Gentile, et al., "Scientific Teaching," *Science* 304, no. 5670 (2004): 521–22.

I only briefly mention "problem-based learning," commonly referred to as PBL, but this brief mention doesn't do service to the long history of this well-developed approach, which has many resources available and a well-established understanding of best practices. A short introduction to PBL can be found at: John R. Savery, "Overview of Problem-Based Learning: Definition and Distinctions," *Interdisciplinary Journal of Problem-Based Learning* 1 (2006): 9–20.

I consider the lecturing versus active learning comparison to be more of a continuum than a dichotomy, because what we consider to be a typical "lecture" class has some active learning elements, even if it is merely asking students to raise their hands once in a while. It is extremely rare for "lecture" or "active learning" courses to occur in their purest form—so from a pragmatic point of view, directly comparing a pure lecture course to a 100% active learning course has little utility to the instructor who is looking to improve their craft. A pragmatic question for most of us, and an answerable question, is "Does the addition of active learning elements to lectures improve student learning?" The answer to this question is an unequivocal "Yes." Here is the definitive meta-analysis demonstrating this in STEM: Scott Freeman, Sarah L. Eddy, Miles McDonough, Michelle K. Smith, Nnadozie Okoroafor, Hannah Jordt, and Mary Pat Wenderoth, "Active Learning Increases Student Performance in Science, Engineering, and Mathematics," *Proceedings of the National Academy of Sciences* 111, no. 23 (2014): 8410–15. This is reinforced by a recent comprehensive report: National Academies of Sciences, Engineering, and Medicine, *How People Learn II: Learners, Contexts, and Cultures* (National Academies Press, 2018).

There is ample evidence that the introduction of active learning strategies improves the educational environment for minoritized students. Studies have shown that active learning helps all students learn better, but the gains for first-generation and ethnic minority students are greater, and the effect sizes in these studies are compellingly large. Two papers I suggest are Julia J. Snyder, Jeremy D. Sloane, Ryan D. P. Dunk, and Jason R. Wiles, "Peer-Led Team Learning Helps Minority Students Succeed," *PLoS Biology* 14, no. 3 (2016): e1002398; and David C. Haak, Janneke Hille Ris Lambers, Emile Pitre, and Scott Freeman, "Increased Structure and Active Learning Reduce the Achievement Gap in Introductory Biology," *Science* 332, no. 6034 (2011): 1213–16.

If you're looking to dip into a bigger "bag of tricks," I highly recommend this briefly annotated list of 280 forms of engagement: K. Yee, "Interactive Techniques," University of San Francisco, 2019, www.usf.edu/atle/documents/handout-interactive-techniques.pdf.

The effect of laptops and mobile phones on student learning in a traditional classroom is a contentious issue. One study from 2014 that gained substantial attention from scholars and the mass media concluded that handwritten notes result in better learning than typing on a laptop: Pam A. Mueller and Daniel M. Oppenheimer, "The Pen Is Mightier Than the Keyboard: Advantages of Longhand over Laptop Note Taking," *Psychological Science* 25, no. 6 (2014): 1159–68. However, a recent replication and extension of this experiment yielded more ambiguous results: Kayla Morehead, John Dunlosky, and Katherine A. Rawson, "How Much Mightier Is the Pen Than the Keyboard for Note Taking? A Replication and Extension of Mueller and Oppenheimer (2014)," *Educational Psychology Review* (2019): 1–28. Other concerns about laptops in the classroom related to student learning (such as accessibility and creating a mutually respectful learning environment) might be a higher priority for some instructors relative to a possible marginal gain experienced by taking written notes in class.

The concept of "learning styles" clearly has won many advocates, within and beyond the educational community. Nonetheless, there is scant empirical evidence to support the notion that learning through an individual's preferred mode (e.g., auditory, visual, textual) results in greater learning, and this is not for lack of investigative effort. The highly cited review article summarizing research on this topic is Harold Pashler, Mark McDaniel, Doug Rohrer, and Robert Bjork, "Learning Styles: Concepts and Evidence," *Psychological Science in the Public Interest* 9, no. 3 (2008): 105–19. It's possible that the propagation of the myth of learning styles is not merely pedagogically neutral but detrimental to student learning, as suggested by a recent study: Shaylene E. Nancekivell, Priti Shah, and Susan A. Gelman, "Maybe They're Born with It, or Maybe It's Experience: Toward a Deeper Understanding of the Learning Style Myth," *Journal of Educational Psychology* (2019): early online.

Chapter 5

A breadth of approaches to assessment are described in this helpful manual: E. F. Barkley and C. H. Major, *Learning Assessment Techniques: A Handbook for College Faculty* (Jossey-Bass, 2016). I've gotten this book as a present for fellow instructors looking for approaches to formative assessment that go beyond clicker questions and short quizzes.

Managing group projects is challenging but often worth the effort. One highly used source for organizing group projects is Elizabeth F. Barkley, K. Patricia Cross, and Claire H. Major, *Collaborative Learning Techniques: A Handbook for College Faculty* (John Wiley & Sons, 2014). Students often complain about difficulties in group work. These student issues are summarized in this short opinion piece: Ann Taylor, "Top 10 Reasons Students Dislike Working in Small Groups . . . and Why I Do It Anyway," *Biochemistry and Molecular Biology Education* 39, no. 3 (2011): 219–20. In this article, Taylor argues, notwithstanding student complaints, "Why do I continue to use small groups in my classroom? Because it works. By the end of the semester, there are improvements in their performance, teamwork, and ability to solve problems. And this is what education is about: students' growth and learning."

Chapter 6

Test anxiety undoubtedly harms the academic performance of students, as well explained in this meta-analysis: Ray Hembree, "Correlates, Causes, Effects, and Treatment of Test Anxiety," *Review of Educational Research* 58, no. 1 (1988): 47–77. If we are successful in developing intrinsic motivation within our students, this will result in greater learning: Edward L. Deci, Robert J. Vallerand, Luc G. Pelletier, and Richard M. Ryan, "Motivation and Education: The Self-Determination Perspective," *Educational Psychologist* 26, no. 3–4 (1991): 325–46.

Multiple-choice questions are simple, but writing a high-quality multiple-choice question is very complicated. Without iterative evaluation of questions, it's impossible to avoid introducing biases unintentionally, and it's very hard to develop questions that test for the specific content you would like to test, without testing for additional or less relevant information. The classic book that is heavily cited is Thomas M. Haladyna, *Developing and Validating Multiple-Choice Test Items* (Routledge, 2004). The brief guidelines that I provided in this chapter were derived from an article by the same expert: T. M. Haladyna, S. M. Downing, and M. C. Rodriguez, "A Review of Multiple-Choice Item-Writing Guidelines for Classroom Assessment," *Applied Measurement in Education* 15 (2002): 309–44, as well as this blog post: C. Brame, "Writing Good Multiple Choice Questions," 2013, https://cft.vanderbilt.edu/guides-sub-pages/writing-good-multiple-choice-test-questions/. While the book is a dense and technical read, the articles are good starting points, and I recommend learning more about writing proper multiple-choice questions if this is a standard mode of testing in your institution.

Chapter 7

Confronted with the mass prevalence of cheating, and the unfairness that this causes for honest students, we have no easy answers. It is no surprise that students are more likely to cheat when they are accustomed to cheating, when they anticipate it will be rewarded, when they think they will not get caught, and when they are not as prepared: Bernard E. Whitley, "Factors Associated with Cheating among College Students: A Review," *Research in Higher Education* 39, no. 3 (1998): 235–74.

Hypervigilance may reduce the incidence of cheating, but it also will harm the educational environment for all students. Two books make a strong case that the cost of draconian prevention measures is not justified by the educational losses: James M. Lang, *Cheating Lessons* (Harvard University Press, 2013); and Donald L. McCabe, Kenneth D. Butterfield, and Linda K. Trevino, *Cheating in College: Why Students Do It and What Educators Can Do about It* (Johns Hopkins University Press, 2012). McCabe et al. focus more on changes at the institutional level that can promote academic integrity and reduce the incidence of cheating, whereas Lang provides very specific suggestions that individual faculty members can adopt to create courses that reduce the incentives for cheating and improve student learning. If you feel that you're struggling with cheating in the classroom, I recommend a close read of Lang. It's not written with science courses in mind, and some of this might not fit what we do in our classes, but it's still helpful for developing a positive instructor mindset.

If you are interested in learning more about the reservations that some professors have about using commercial plagiarism-detection software, I recommend reading: S. M. Morris and J. Stommel, "A Guide for Resisting EdTech: The Case against Turnitin," 2017, http://hybridpedagogy.org/resisting-edtech/.

Chapter 8

One highly cited guide to the practical matters of teaching an online course is: Susan Ko and Steve Rossen, *Teaching Online: A Practical Guide* (Taylor & Francis, 2017). If you're looking for a handy text specifically to guide you through the details of developing and running an online course, this will be a valuable resource. Another volume that is full of practical advice, with more theoretical context, is: Claire Howell Major, *Teaching Online: A Guide to Theory, Research, and Practice* (Johns Hopkins University Press, 2015).

Online learning happens when students and their instructor form what is often referred to as a Community of Inquiry. According to this theory, the community has three aspects: social presence, cognitive presence, and teaching presence. The Community of Inquiry framework is useful in the context of online teaching because intentional design is typically needed to foster well-developed communities focused on learning. The landmark publication building this framework is: D. Randy Garrison, Terry Anderson, and Walter Archer, "Critical Inquiry in a Text-Based Environment: Computer Conferencing in Higher Education," *The Internet and Higher Education* 2, no. 2–3 (1999): 87–105. The way that a professor engages with their students, and how students feel their teaching presence, is connected to student satisfaction and student learning. See: Richard Ladyshewsky, "Instructor Presence in Online Courses and Student Satisfaction," *International Journal for the Scholarship of Teaching and Learning* 7, no. 1 (2013): 1–23.

While there doesn't seem to be much evidence that including more instructional videos improves student learning, students feel positive about them. This topic is reviewed here: Peter J. Draus, Michael J. Curran, and Melinda S. Trempus, "The Influence of Instructor-Generated Video Content on Student Satisfaction with and Engagement in Asynchronous Online Classes," *Journal of Online Learning and Teaching* 10, no. 2 (2014): 240–54. This study also suggests that it's more impactful when the video content is created by their own instructor.

Discussions of educational equity for students with marginalized identities recognize the pitfalls within the attempts to provide access for all students. In many studies, educational disparities between white students and students of color are amplified in online courses. Nevertheless, online courses are often promoted as a mechanism to provide access to educational opportunities for traditionally underrepresented students. This doesn't mean that online teaching is inherently inequitable, but instead, that all instructors need to have an equity-first mindset while teaching and adopt practices for equitable course design. This need is discussed in this study: Angelica M. G. Palacios and J. Luke Wood, "Is Online Learning the Silver Bullet for Men of Color? An Institutional-Level Analysis of the California Community College System," *Community College Journal of Research and Practice* 40, no. 8 (2016): 643–55.

A short article that provides more practical tips and helpful links about moving laboratories online in a crisis is from Heather R. Taft: https://www.chronicle.com/article/How-to-Quickly-and-Safely/248261.

One research study about students accepting a 24-hour response time to emails is: Ching-Wen Zhang, Beth Hurst, and Annice McLean, "How Fast Is Fast Enough? Education Students' Perceptions of Email Response Time in Online Courses," *Journal of Educational Technology Development and Exchange* 9, no. 1 (2016): 1. Based on their survey results, students communicated what I think are reasonable expectations—they don't expect us to respond to emails during evenings, weekends, or holidays, and as long as they hear back the next business day, most students are okay with that time lag.

Suggested Readings

If you're ready to read one more book to learn more about how to teach effectively, I suggest: Susan A. Ambrose, Michael W. Bridges, Michele DiPietro, Marsha C. Lovett, and Marie K. Norman, *How Learning Works: Seven Research-Based Principles for Smart Teaching* (John Wiley & Sons, 2010).

Although written with K–12 teachers in mind, I think there are a large number of straightforward lessons in this book that can help you find a way to improve your craft: Robyn R. Jackson, *Never Work Harder Than Your Students and Other Principles of Great Teaching* (ASCD, 2018).

A great set of lessons and anecdotes can be found in this classic, even though it's not directed at the sciences: Ken Bain, *What the Best College Teachers Do* (Harvard University Press, 2011).

An engaging combination of evidence and stories, with respect for students at the forefront, is Joshua R. Eyler, *How Humans Learn: The Science and Stories behind Effective College Teaching* (West Virginia University Press, 2018).

As scientists, we regularly work with students to create new knowledge in the lab and in the field. This landmark volume presents the argument that we must regard our students as fellow investigators and that we are working alongside them to learn: Paulo Freire, *Pedagogy of the Oppressed* (New York: Herder and Herder, 1972).

Index

academic honesty, 155
academic misconduct, 161; cheating, 151–53; honor code, 150–51; plagiarism, 152–53, 158–59; Respect Principle, 152
active learning, 94, 167; clickers, 92; as effective, 77–79; as exhausting, 81; flipping a course, 75; formative assessment, 98; vs. lecturing, 75–76, 80, 170; minoritized students, 171; problem-based learning, 75; team-based learning, 75
active shooters, 139–40
Adams, Eldridge, 87–88, 94, 163, 179
American Chemical Society (ACS), 49–50
assignments: homework, 105–6; extra-credit assignments, 113–14; grading rubrics, 106–8; group projects, 108–10, 172; lab assignments, 115–16; and plagiarism, 111–13; scaffolding writing, 110–11, 113
Association of College and University Educators (ACUE), 164–65
asynchronous instruction, 168; benefits of, 166; flexibility of, 166; office hours, 174; as self-paced, 165
Ayala, Eduardo, 15

Bloom's Taxonomy, 20, 58, 62, 120, 125, 131, 168
Boyle, Alice, 36

Calendly, 174
California State University system, 8, 167–68
Center for Teaching and Learning, 134, 148, 164
Chakrabarty, Prosanta, 17, 19
cheating, 168; academic misconduct, 151–53; anxiety, 153–54;

cheating, 168 (*cont.*) consequences of, 154–55; hypervigilance, 173; in-class, 154–55; minimizing, on exams, 153–54; testing, 175
classroom management, 170
classroom response systems, 91–92. *See also* clickers
clickers, 97, 148; as active learning technique, 92; Efficient Teaching, 91; gripes about, 92. *See also* classroom response systems
common problems: academic misconduct, 150–53; authority, disrespect for, 140–42; catching problems early, 133–34; cheating on exams, 153–54; disrespectful behavior, 140–41; disruptive or threatening behavior by students, 139–40; finding a mentor, 134; grade change requests, 134–36; handling plagiarism, 158–61; in-class cheating, 154–55; low attendance, 147–48; minimizing plagiarism, 155–58; monopolizing discussions, by students, 143–44; sexual misconduct, 142–43; soliciting anonymous feedback, 134; student overentitlement, 144–46; student privacy, 136–39; student underentitlement, 146–47
Community of Inquiry: aspects of, 189–90; cognitive presence, 190; social presence, 190; teaching presence, 190
COVID-19 pandemic, 162, 176
curriculum, 49; backward design, 50–51, 57–58, 66; choosing textbooks, 60–61; covering material vs. teaching content, 52–55; culturally responsive teaching, 63–64; depth vs. breadth, 52, 169; developing a course from scratch, 49; difficulty setting, 61–62; discipline-specific standards, 169; information literacy, 62–63; learning objectives, 55–56; lesson plans, 64–65; memorization, 58–60; teaching evaluations, 65–69

Darwin, Charles, 63
Deepwater Horizon, 64
DeVaney, Shannon, 63, 163, 179
digital devices: banning of, 93–95; as distraction, 93; laissez-faire approach to, 94–95; pragmatic approach to, 95
digital learning, 97
digital natives, 62
disabled students, 94
Disabled Student Services, 44–45
disrespectful behavior: categories of, 140–41; graduate teaching assistants, 141; new faculty, 141
Douglass, Frederick, 167
Dunning-Kruger effect, 143

educational equity: marginalized identities, 190
effective learning, 55
Efficient Teaching, 2–4, 66–67, 74, 95, 106; cheating, 152; clickers, 91; online teaching, 162
emails: office hours, 174; as positive, 171; response time to, 172–73, 190
equitable teaching, 171; access to devices, 169–70; marginalized students, proactive toward, 170, 190; roadmap for success, 170; student diversity, 169
evaluations. *See* teaching evaluations
exams: cheating, 175; cooperative, 132; criteria-based grading, 131; difficulty level, setting of, 119–20; evaluation biases, 118–19;

expectations of, 124; Fairness Principle, 124; fear of, 123; fielding student questions, 127–28; final, 121; as formative assessments, 118, 132; grading of, 128–29; group exams, 118; handing back, 129; honest behavior, expectations of, 175; kinds of, 120–21; learning disabilities, 118–19; as learning experiences, 131; Learning Management System (LMS), 127, 129–31; logistic difficulties of, 118; midterm, 121–22; as motivator, 118; multiple-choice questions, 10, 92, 97, 122, 172; old exams, copies of, 129–30; in online teaching, 174–76; open-book, 120–21, 175; positive feedback loop, 127; questions, reusability of, 131; regrading of, 131–32; retaking of, 132; review sheets, 124–26; Scantron forms, 130; and stress, 124; as summative assessments, 117–18, 131; surveillance, recommendations against, 175; take-home, 121, 175; test anxiety, 119, 172; test banks, 130; traditional, 118, 120; writing good exam questions, 122–23

faculty-student relationships, 12
Fairness Principle, 124
Family Educational Rights and Privacy Act (FERPA), 136–38
Finkelman, Leonard, 37
food insecurity, 8, 167
formative assessments: active learning, 98; disadvantages of, 97–98; as effective pedagogy, 95–96; forms of, 96; mobile polling, 97; online portals for digital learning, 97; quizzes and homework, 96; specific and prompt instructor feedback, 96

Franklin, Rosalind, 63
Freeman, Nancy, 41

gaming the system, 11–12
Gates, Bill, 167
Gerardo, Nicole, 19
Gilbert, Neil, 15, 90
grading, 168; on a curve, 40; performance benchmarks, 41. *See also* assignments; exams; syllabus
grading rubrics, 110, 128–29, 154; lab reports, 116; as powerful tool, 108
grading system, 107; rubrics and teaching efficiency, 107–8
group projects, 108, 172; team-based learning approach, 109

Heil, Martin, 163, 179
higher education institutions, 21; community colleges, 23; performance in, 66; small liberal arts colleges (SLACs), 22–23; research-intensive institutions, 22; teaching-focused universities, 22
homelessness, 8, 167
Hörst, Sarah, 95

Lacks, Henrietta, 64
Lang, James M., 173
Lawson, Sarah, 109, 162, 178
learning, 178; handwritten notes, 171; kinds of, 20; laptops, effect on, 171
Learning Management System (LMS), 148, 165–66, 169–70, 174–75; benefits of, 89–90; course content, questions about, 171–72; drawbacks of, 89; discussion board, 168; exams, 127, 129–31; online forum, 90; plagiarism, 112; student privacy, 137, 164; syllabus, 44; teaching methods, 89–91, 97
learning outcomes, 19–20

learning styles, 171
lecturing, 77, 79; active learning, alternative to, 75–76, 80, 170
Letcher, Susan, 69, 162, 178

managing communications: emails, 171–72
Margulis, Lynn, 63
McCabe, Donald L., 173
McCreery, Helen, 19, 163, 179
memorization, 58, 60; "cheat sheet," 59, 169; as ritual hazing, 58–59
Mendel, Gregor, 63
metacognition, 84–85
Mohs scale, 20–21
MOOCs (massive open online courses), 162

office hours, 173; asynchronous instruction, 174; different kinds of media, 174; emails, 174; syllabus, 43–44; teaching methods, 85–87; third-party scheduling apps, 174
online teaching: asynchronous instruction, 165–66, 168; "breakout rooms," 168; characteristics of, 164–65; Community of Inquiry, 189–90; during COVID-19 pandemic, 162, 176; creating community, 167–69; as different, 163, 165; digital work, incorporating of, 163–64; drawbacks of, 166–67; Efficient Teaching, 162; emails, 171–72, 190; equitable course design, 169–71, 190; flexibility of, 163, 166; group learning, well suited to, 167; "icebreaker" activities, 168; instructional videos, 190; introductory videos, sharing of, 167–68; isolation, 167; keeping students on track, 166–67; laboratories, 176–77; managing communications, 171–72; notions of, as misleading, 162; office hours, 173–74; "participation" points, 168–69; Respect Principle, 162; student diversity, 169; student privacy, 164; students, from marginalized backgrounds, 170, 190; synchronous instruction, 165–66; synchronous video sessions, 173; testing, 174–76; upside of, 163. *See also* MOOCs (massive open online courses)
OpenStax, 61
Ory, Jeramia, 124

pedagogical content knowledge (PCK), 50
plagiarism, 111, 168; academic misconduct, 152–53, 158–59; asynchronous instruction, 165; contract plagiarism, 113, 157; denial of, 159–60; detection software, 112; handling of, 158–61; intentional vs. unintentional, 156; Learning Management System (LMS), 112; levying sanctions for, 161; minimizing of, 155, 156, 157, 158; Respect Principle, 112, 158; scaffolding written assignments, 113, 158; syllabus, 46, 47, 157
plagiarism-screening services: detection, 112–13, 173; prevention, 112–13
problem-based learning (PBL), 170

Quality Matters (QM), 164–65

Respect Principle, 2, 4–5, 66, 72, 95, 106, 134–35, 149; academic misconduct, 152; educational equity, 169; extra-credit assignments, 114; online teaching, 162; and plagiarism, 112, 158; students' lives, complexity of, 6–9; student overentitlement, 144; students monopolizing discussion, 143; treating students fairly, 6, 9–11

Salinas, Santiago, 86
scaffolded writing assignments: and plagiarism, 158
science communication: deficit model of, 24
science dismissers, 148, 150; Respect Principle, 149
scientific ideas, 63
Scientific Teaching, 170
sexual assault, 142–43
sexual harassment, 142
Stommel, Jesse, 112
student deference: toward teachers, 16–17
student engagement, 65, 92, 94
student learning: responsibility for, 14–15
student overentitlement, 144; as distraction, 145; extra credit, 145; points, 145; shaming behaviors, 145–46; time, 145
student privacy, 139; Family Educational Rights and Privacy Act (FERPA), 136–38; Learning Management System (LMS), 137; online teaching, 164
student underentitlement, 147; as pervasive problem, 146
syllabus, 95, 114; academic misconduct, 45–47, 151; accommodations for disabilities, 44–45; activities outside classroom, 45; audio or video recording, 42; campus resources, 47–48; cheating, 45–46, 154; contact expectations, 43; curve grades, 40; dealing with problems, 26–27; disruptive students, 42; efficiency, 26; expected learning outcomes, 31–32; extra credit, 39; grade appeal, 27; grade distribution, 37–38; grading policies, 34–35; importance of, 26–27; lab policies, 45; late assignments, 36–37; Learning Management System (LMS), 44; as legal document, 26, 29–30, 48; letter grade calculation, 40–41; missed classes and exams, 35–36; office hours, 43–44; participation score, 38–39; personal travel, 41–42; phone and laptop use, 41; plagiarism, 46–47, 157; point system, 37–38; as powerful, 27; problem students, 26; required materials, 31; schedule of classes, 32–33; set of behavioral norms, 28–29; as set of rules, 26–27; standard format, 30–32; student excuses, 33–35; study tips, 47; as template for future semesters, 48; tone for semester, 28; university-required elements, 44; updating of, 48
synchronous instruction: community of learners, building of, 165–66

Taylor, Ann, 172
Tharp, Marie, 63
teachers: dress of, 17–18; responsibility of, 16
teaching, 178; as art, 165; as series of decisions, 1
teaching-as-coaching approach, 12–13
teaching assistants, 22, 96; lab section, 141
teaching evaluations, 65–66; administering of, 67–68; approachability, impact of, 69; biases in, 69, 169; likability, impact of, 69; supplemental evaluation forms, 68

teaching methods: active learning, 75–81, 92, 94, 98; asking questions, 92–93; bag of tricks, 82–83, 171; best practice, 70; case studies, 83; classroom management, 71–73; classroom response systems, 91–92; class time and student learning, 74–75; consistency of science classes, 71; digital devices, 93–95; evidence-based teaching practices, 74; formative assessment, 95–98; getting buy-in from students, 84; information delivery, 99; lab and field safety, 99–100; lab attendance, 103–4; lab partnering, 102–3; lab sessions, teaching of, 100–102; Learning Management System (LMS), 89–91, 97; learning styles, 98–99; lecturing, 75–77, 79; metacognition, 84–85; office hours, 85–87; scientific teaching, 73–74; slides, 87–89; think-pair-share, 82
teaching philosophy, 18–19, 25
teaching styles, 13; being yourself, 14; empathy, 14
terHorst, Casey, 85
tests. *See* exams
Title IX, 142–43

United States, 21, 136, 139–40, 142; diversity in, 63
university campuses: sexual assault on, 142

Venesky, Matthew, 121–22

Wegener, Alfred, 63
Williamson, Jessie, 17, 163, 179

Zoom, 173

Made in the USA
Columbia, SC
16 November 2020